Environmental Injustice and Catastrophe

De Gruyter Contemporary Social Sciences

Volume 24

Environmental Injustice and Catastrophe

How Global Insecurities Threaten
the Future of Humanity

Edited by
Baris Cayli Messina

DE GRUYTER

ISBN 978-3-11-162922-3
e-ISBN (PDF) 978-3-11-108168-7
e-ISBN (EPUB) 978-3-11-108209-7
ISSN 2747-5689
e-ISSN 2747-5697

Library of Congress Control Number: 2023930110

Bibliographic information published by the Deutsche Nationalbibliothek
The Deutsche Nationalbibliothek lists this publication in the Deutsche Nationalbibliografie; detailed bibliographic data are available on the Internet at http://dnb.dnb.de.

www.degruyter.com

Table of Contents

Baris Cayli Messina
Introduction

We are all susceptible to the repercussions of risk in this increasingly interconnected and globalized society. As a result of scientific and technological progress, we
have been able to improve the quality of our lives and better defend ourselves
from potential threats to our health and safety. At the same time, however, we become more fragile and are exposed to a multitude of risks. From nuclear disaster
risks to pandemic, we begin to consider our future, which induces stress and anxiety about how to protect ourselves and our loved ones, as well as how to live in a
sustainable and peaceful society. Giddens defined a risk society as one in which
concerns about the future determine fundamental safety measures and cause individuals to reconsider the impact of risk on their daily lives.[1] On the other hand,
Beck linked the growing impact of risks to the insecurities and threats that were
systematically created and unleashed by the modernization process.[2] He defined
modernization in the context of significant changes influencing our behaviors as
a result of technological rationalization and advancements in work and organization, and he cautioned that modernization brings more rapid and significant transformations, arguing that broad risks of modernization have a "boomerang effect"
in which those who generate the risk are also vulnerable to its consequences. The
most obvious manifestation of this boomerang effect, which endangers the future
of humanity, is the destructive impact of climate change. Therefore, environmental
injustice results in more detrimental consequences, necessitating urgent questions
and investigations to rethink the harms to individuals, organizations, and states,
while identifying those sinister networks and actions that severely risk the future
of humanity.

Certain societies have historically been more prone to natural disasters due to
the terrain in which they reside, so much so that they have developed a disaster
culture in which natural disasters have altered the cultural fabric of those communities.[3] Despite the scientific and technological advancements of the last two
centuries, both the developed and developing worlds are vulnerable to risks, albeit
to varying degrees. However, geographical risks and environmental hazards only
provide us with a partial picture. Some communities live in limbo, accumulating

---

1 Anthony Giddens, *Consequences of Modernity*. Cambridge, England: Polity Press, 1990.
2 Ulrich Beck, *Risk Society: Towards a New Modernity*. Translated by Ritter, Mark. London: Sage
Publications, 1992.
3 Gregg Bankoff, *Cultures of Disaster: Society and Natural Hazards in the Philippines*. London:
Routledge, 2003.

https://doi.org/10.1515/9783111081687-001. The original version of this chapter has been revised. Unfortunately, the
name of a contributor was misspelled in the original publication due to a production error. This has been corrected,
along with some typographical errors. The press apologizes for any inconvenience caused.

insecurities as a result of the uncontrolled modernization process in which the free market and corporations prioritize profits over the welfare of communities, sustainable environment, and a safe future. Furthermore, the lack of a common platform where domestic interests align with international environmental policy-making prevents long-term solutions from being implemented globally.[4]

The purpose of this book is to theorize how human activities contribute to risks and insecurities that threaten environmental sustainability, the future of our planet, and human security. By situating the environmental crisis within a broader interdisciplinary context and bridging the humanities and social sciences, we seek to identify the factors and actors that contribute to environmental injustice, whose catastrophic effects are felt by the most vulnerable communities. The environmental crisis struck the world at the turn of the twentieth century was unprecedented in its scope, severity, and speed.[5] The problem has only deteriorated over the last two decades. Environmental disaster was accelerated by anthropogenic climate change, declining air and water quality, land contamination due to chemical and radioactive pollution, deforestation, soil erosion, habitat and biodiversity loss, wars, and escalating political strife. By centralizing human activities and identifying the causes and effects of environmental risks and disasters in this book, we uncover various forms of human activity that shape the scale of global insecurities and disasters with severe repercussions. In doing so, we hope to unveil the relationship between detrimental human activity and global insecurities, which threatens our the existence of human being and throws our future into disarray. Different chapters in this book aim to convey the message of how and to what extent we are destroying our planet and the dire consequences of environmental injustice and its catastrophes. Public ire and collective rage are fueled by the absence of a functional social state, the lack of global governance of environmental risks, and sufficient political capacity to protect the rights of vulnerable individuals, dejected communities, and the future of our environment. As a result, we face a variety of threats and acts of violence that disproportionately affect certain regions and populations. Theorizing global insecurities in this age of environmental crises, the purpose of this book is to shed new light on the fragility of our planet, where different violence forms and human activities threaten our existence and imperil the future of humanity.

---

4 Dana R. Fisher, *National Governance and the Global Climate Change Regime.* New York: Rowman & Littlefield, 2004.
5 Chris C. Park, *The Environment: Principles and Applications.* London and New York: Routledge, 2001.

Theorizing Global Insecurities in the Age of Environmental Crisis

In a world dominated by violence, escalating conflict heightens risks and insecurities. Political solutions to such threats encouraged ideological experiments that shaped government forms and everyday life. Some early 20th century thinkers, like the French philosopher Georges Sorel, encouraged collective violence on the grounds that the proletariat needed it to spark a revolution.[6] Social engineers like Mussolini were drawn to "creative violence" because he believed it would thwart socialism and pave the way to a better future for Italy.[7] His fascist ideology created mass killings and a collective trauma for future generations. The connection between political ideologies/political conflict and environmental impact has been mostly neglected. Yet social and political theorists developed a focus on conflict and violence that was profoundly influenced by environmental factors and the human-environment relationship. For example, Karl Marx's social theory has its roots in the interplay between humans and their environments; this interaction was crucial in the development of the social metabolism of capitalist societies, and specifically of industrial agriculture.[8] By diminishing the productivity of land and generating non-recyclable waste in metropolitan areas, capitalist-oriented agriculture had a detrimental effect on the health of urban residents.[9]

The fetishizing of urban violence in the field of geography and social sciences overlooked the role of socio-spatial relations.[10] The urban insurgency opposes policies that exacerbate urban insecurity, demonstrating that climate mobilization is inherently intersectional, and the urban dimension of the global climate insurgency undermines the gap between rural and urban through grassroots movements worldwide.[11] With the rise of urbanization, not only cities but also their outskirts face the threat of poverty, social and environmental degradation. With the rapid industrialization and migration to cities over the last century, this has become a

---

6 Georges Sorel, *Reflections on Violence* (1915). Translated by T. E. Hulme. New York: Peter Smith, 1941, 177.

7 Zeev Sternhell, Mario Sznajder, and Maia Ashéri, *The Birth of Fascist Ideology: From Cultural Rebellion to Political Revolution.* Princeton, NJ: Princeton University Press, 1994, 93.

8 John Bellamy Foster, "Marx's Theory of Metabolic Rift: Classical Foundations for Environmental Sociology". *The American Journal of Sociology* 105 (1999): 366–405.

9 Foster, "Marx's Theory of Metabolic Rift", 1999, 378.

10 James Tyner and Joshua Inwood, "Violence as fetish: geography, Marxism, and dialectics". *Progress in Human Geography* 38 (2014): 771–784.

11 Ashley Dawson, Marco Armiero, Ethemcan Turhan, and Roberta Biasillo, "Urban Climate Insurgency: An Introduction". *Social Text* 40 (2022): 1–20.

recurring and global issue.[12] The gentrification of cities increases business development, but it also creates suburbs by deepening the social precariousness of certain communities.[13] Unsafe and unequal economic distribution make rural areas insecure, whereas urban sprawls and ghettos produce a different form of insecurity characterized by accelerated violence under the control of gangs, mafias, or the police state.[14] The rich literature on conflict and violence in urban social space offers us a powerful conceptual matrix to explore different and complex factors that play role in shaping spatial context in the urban space.[15] Besides, a chaotic political environment provides new opportunities to extra-legal groups in which political violence may swiftly transform into criminal violence.[16] The control of extra-legal groups from the construction industry to waste management by means of corruption and patronage-client relationships creates additional environmental risks and hazards.

The relationship between global insecurities and environmental crisis cannot be separated from power hierarchies and relationships. The power dynamic that exists between various social agencies determines how society's foundational norms, which have an impact on the most important aspects of community life, are governed.[17] Examining hierarchal power relationships exposes asymmetrical social networks through which domination determines the governance of relations and diminishes the influence of struggles for equal access to opportunities. As Foucault stated, "Domination is in fact a general structure of power whose ramifications and consequences can sometimes be found descending to the most recalcitrant fibers of society."[18] An investigation into the hierarchical power relationship sheds light on the distribution of power in society as well as the ways in which it influences the social and political structure, the movement of goods,

12 Caroline O. N. Moser, "Urban violence and insecurity: an introductory roadmap". *Environment and urbanization* 16 (2004): 3–16. See also Martin Coward, "Network-centric violence, critical infrastructure and the urbanization of security". *Security dialogue* 40 (2009): 399–418.
13 Lanre Davies, "Gentrification in Lagos, 1929–1990". *Urban History* 45 (2018): 712–732.
14 Roberto Aspholm, *Views from the Streets: The Transformation of Gangs and Violence on Chicago's South Side.* New York: Columbia University Press, 2020.
15 Marion Neale, "Research in urban history: Recent theses on crime in the city, 1750–1900". *Urban History* 40 (2013): 567–577.
16 Mortiz Schuberth, "A transformation from political to criminal violence? Politics, Organised crime and the shifting functions of Haiti's urban armed groups". *Conflict, Security & Development* 15 (2015): 169–196.
17 Manuel Castells, "A Sociology of Power: My Intellectual Journey". *Annual Review of Sociology* 42 (2016): 1.
18 Michael Foucault, "The subject and power". *Critical inquiry* 8 (1982): 795.

and the social networks.[19] Such investigation assists us uncovering the origins of conflicts which have a greater role in the surge of global insecurities.

By shifting our attention to conflicts, we can explain how the distribution of power regulates relationships between different actors and leads to catastrophic events that destroy the environment and human lives. Social conflicts bring direct violence to the center of everyday life, which can be seen from urban riots to migrations to cities.[20] On the other hand, when we examine political conflict, we can determine how resources are used to make matters worse between competing state actors. This can result in environmental catastrophes and major threats to humanity, such as the Cuban Missile Crisis and the nuclear war threat during the Cold War. John Keane contends that the primary factor that contributes to the success of political incompetence is the failure of suppressed publics.[21] Therefore, the catastrophic occurrences would not occur if the apathy of the populace was not common. 'Carbon filtration plants, undersea tunnels, high-speed railway networks, new airports and airport extensions, research and development of new weapon systems, liquid natural gas plants, new communications systems, and nuclear power plants' are examples of state- or corporation-funded megaprojects that carry a high risk of leading to a catastrophe.[22] Keane's argument finds a similar echo in Bernstein's claim that if people unite around similar and non-violent goals, they can prevent autocratic and violent governance from destroying the environment and suppressing communities.[23] Don Grant, Andrew Jorgenson, and Wesley Longhofer's recent sociological research on carbon dioxide emissions from power plants reveals that structural concerns must be remedied by developing effective regulatory mechanisms, improving technical efficiency, and encouraging public involvement to reduce pollution and environmental risks.[24] Yet the poignant reminders is that while there are encouraging examples to draw from, the vast majority of people in the world do not have the means to effectively challenge their governments and

---

19 Michael Mann, *Sources of Social Power, Volume 1*. Cambridge: Cambridge University Press, 1986.
20 For urban riots see, Arnaud Exbalin, "Riot in Mexico City: A challenge to the colonial order?" *Urban History* 43 (2016): 215–231. For migration into cities and social conflict see, Erika Hanna and Richard Butler, "Irish urban history: An agenda". *Urban History* 46 (2019): 2–9.
21 John *Keane, Violence and Democracy. Cambridge, UK: Cambridge University Press, 2004.*
22 John Keane, "Silence and Catastrophe: New Reasons Why Politics Matters in the Early Years of the Twenty-first Century". *The Political Quarterly* 83 (2012): 660–668.
23 Richard J. Bernstein, *Violence: Thinking without Banisters*. Cambridge: Polity Press, 2013.
24 Don Grant, Andrew Jorgenson, and Wesley Longhofer, *Supper Polluters. Tackling the World's Largest Sites of Climate-Disrupting Emissions*. New York: Columbia University Press, 2020.

Disasters and Environment in Our Fragile World

In the last century, technology has accelerated the visibility of catastrophic incidents, and the media has amplified their global impact through the transmission of devastating photographs, documentaries, and videos. Thanks to the media outlets, we are now able to comprehend the disastrous effects of violence and the principal actors who attempted to justify invasions, wars, and the experiment of heavy armaments over the civilian population which created great environmental hazards at the same time. Human activity causes population decline, which leads to the collapse of civilization due to environmental destruction, hostile neighbors, and adversarial social interactions.[25] Globalization under the control of imperial powers manifested its failure through bloody conflicts that we experienced, including the American invasion of Iraq and the recent Russian invasion of Ukraine.

There may not appear to be much of a distinction between natural hazards and disasters at first glance. The key difference is that disasters serve as sobering reminders of the consequences of human action. The impact of climate change on various natural disasters demonstrates that humans are usually to blame for environmental hazards.[26] Climate change and human-made natural disasters are inextricably linked.[27] Through liberal policies characterized by exponential energy consumption and resource exploitation, primarily by the wealthy and powerful nations of the global North, globalization manifested its most severe consequences. More than 1,6 million wildfires broke out across the United States between 1992 and 2015. And 96 percent of the fires that came close enough to threaten people's houses were sparked by human activity as opposed to natural causes.[28]

Corrupt governmental regimes and unsustainable forms of economic regulations globally pose serious risks to the vulnerable groups and prevent them accessing to safe place and sufficient food. We can distinguish how human originated problems contribute to the severe impact natural hazards in times of social tragedy and unprepared natural hazards as it happened after the earthquake in Chile

---

25 Jered Diamond, *Collapse: How Societies Choose to Fail or Succeed.* New York: Viking Press, 2005.
26 Phil O'Keefe, Ken Westgate, and Ben Wisner, "Taking the naturalness out of natural disasters". *Nature* 260 (1976): 566–567.
27 Bill McGuire, *Walking the Giant: How a changing climate triggers earthquakes, tsunamis, and volcanoes.* Oxford: Oxford University Press, 2012.
28 Sarah DeWeerdt, "Humans cause 96% of wildfires that threaten homes in the U.S". *Anthropocene* 2020. See also, Nathan Mietkiewicz, Jennifer K. Balch, Tania Schoennagel, Stefan Leyk, Lise A. St. Denis, and Bethany A. Bradley, "In the line of fire: Consequences of human-ignited wildfires to homes in the U.S. (1992–2015)". *Fire* 2020.

in 2010.[29] The fast rise of the world's population and its growing concentration in perilous regions have exacerbated the severity of natural disasters. A changing climate and unstable landforms, as well as deforestation, make disaster-prone areas more vulnerable. And developing nations pay a steep price for these risks.[30]

Wars and violent conflicts between nations cause devastation and leave future generations with a difficult legacy. The likelihood of a catastrophic event is significantly increased by military tensions and the arms race. We had to bear witness to the deaths of thousands of people and the environmental devastation caused by the atomic bombings of Hiroshima and Nagasaki on August 6 and 9, 1945.[31] The future of humanity is threatened by nuclear weaponry and energy dependence. On a local level, these extremely dangerous establishments pose grave risks to the surrounding community. The Hanford Site is one of these establishments which was built in 1943 as a decommissioned nuclear manufacturing facility maintained by the federal government of the United States on the Columbia River in Benton County, Washington. The location has also been referred to as the Hanford Project. This is one of the numerous projects demonstrating how state institutions threaten human and environmental safety. On the construction site, pump systems extracted river water for cooling, cleaned it for use in the reactors, and then returned it to the river. Before being discharged into the river, the spent water was kept in massive containers known as retention basins for up to six hours. This retention had no effect on isotopes with longer half-lives, and the river continued to fill with terabecquerels every day. The federal government withheld information regarding these radioactive discharges polluting environment from 1944 to 1971.[32] Radiation was eventually recorded as far west as the Washington and Oregon beaches, 320 kilometers downstream.[33] Since 2003, radioactive materials are known to be leaking from Hanford into the environment. Those who resided downwind from Hanford or used the Columbia River downstream were exposed to heightened radiation levels that enhanced their risk for many cancers and other disorders.[34] A class action case brought against the federal government by 2,000 Hanford down-

---

29 Keving A. Gould, M. Magdalena Garcia, and Jacob A. C. Remes, "Beyond 'natural-disasters-are-not-natural': the work of state and nature after the 2010 earthquake in Chile". *Journal of Political Ecology* 23 (2016): 93.
30 Adedeji, Daramola and Eziyi O. Ibem, "Urban environmental problems in Nigeria: Implications for sustainable development". *Journal of Sustainable Development in Africa* 12 (2010): 124–145.
31 Kyoko Iriye Selden and Mark Selden, *The Atomic Bomb: Voices from Hiroshima and Nagasaki*. New York: Taylor & Francis, 2015.
32 "An Overview of Hanford and Radiation Health Effects". *Hanford Health Information Network*, accessed on 20 May 2022.
33 Laura Fermi, *The Story of Atomic Energy*. New York: Random House, 1961, 113.
34 "An Overview of Hanford and Radiation Health Effects", accessed on 21 May 2020.

winders lasted several years in court. The Department of Energy concluded the final cases in October 2015. They paid over $70 million in attorney fees and $7 million in damages.[35] The compensation, despite the lengthy court proceedings, might sound to be a positive result. Yet there is no guarantee that this would not repeat again. In many developing nations, environmental crimes committed by states or corporations go unpunished and millions of people become victims of a corrupt government that threatens ecosystem and human security.

The Chernobyl Disaster happened on April 26, 1986 at the No. 4 reactor of the Chernobyl Nuclear Power Plant, which located close to the city of Pripyat in the northern Ukrainian SSR in the Soviet Union.[36] The reactor was on the verge of exploding due to a combination of operator negligence and serious design flaws. Instead of stopping the test, an uncontrollable nuclear chain reaction began, releasing massive amounts of energy.[37] The Chernobyl disaster was considered to be the worst nuclear disaster in the history, along with the nuclear disaster that took place in Fukushima, Japan in 2011.[38] The nuclear disaster at Chernobyl caused an environmental catastrophe and presented a significant threat to public health in the region affecting more than a dozen of country directly.[39] Subsequent study on microorganisms indicated that bacterial and viral specimens exposed to the radiation (including Mycobacterium tuberculosis, herpesvirus, cytomegalovirus, and tobacco mosaic virus) underwent fast alterations shortly after the disaster.[40]

Disasters have a direct impact on the environment, but environmental damage does not always receive the same attention as human loss, and the role of social class in environmental disasters is frequently overlooked. This was painfully demonstrated by Hurricane Katrina. In late August 2005, Hurricane Katrina slammed New Orleans and the surrounding region, killing over 1,800 people and inflicting $125 billion in damages. The floodwaters also devastated the majority of New Orleans' transportation and communication networks. Consequently, tens of thou-

---

35 Rebecca Boyle, "Greetings from Isotopia". *Distillations* 3 (2017): 26–35.

36 Peter Burgherr and Stefan Hirschberg, "A Comparative Analysis of Accident Risks in Fossil, Hydro, and Nuclear Energy Chains". *Human and Ecological Risk Assessment* 14 (2008): 947–973.

37 "The Chernobyl Accident: Updating of Insag-1 Insag-7", *A report by the International Nuclear Safety Advisory Group*, Vienna: International Atomic Agency, 1992.

38 Peter van Ness and Mel Gurtov (eds.), *Learning from Fukushima: Nuclear power in East Asia.* Canberra, Australia: ANU Press, 2017.

39 "Environmental Consequences of the Chernobyl Accident and their Remediation: Twenty Years of Experience", *Report of the Chernobyl Forum Expert Group 'Environment'.* Vienna: International Atomic Energy Agency (IAEA), 2006, 23–25.

40 Alexey Vyablokov, Vassily B. Nesterenko, and Alexey V. Nesterenko, "Consequences of the Chernobyl Catastrophe for the Environment". *Annals of the New York Academy of Sciences* 1181 (2009): 221–286.

sands of individuals who had not evacuated the city before to the hurricane were left without food, shelter, and other needs. After Hurricane Katrina, the trend of increasing danger for the poor, ethnic minorities, and other disadvantaged members of society re-emerged as the primary concern. As a significant portion of the population impacted by Hurricane Katrina had the lowest socioeconomic status, the connections between high-risk populations and day-to-day health disparities were glaringly obvious.[41] African Americans were significantly more likely than White Americans to express rage and depression in response to the hurricane's events. In addition, the majority of these emotions derive from the belief that Hurricane Katrina, from its impact to its media coverage and government response, was perceived as a racially motivated occurrence.[42]

People and communities have lost their homes, towns, landscapes, and ways of life due to rising global temperatures, and these losses are compounded by the fact that their insurance coverage and housing investments have been destroyed or rendered useless by the ensuing floods and storms.[43] In a socially stratified society, the lives of those at the bottom of the social hierarchy are negatively impacted by the severity of ecological disasters. Without coordinated action, as many as 325 million people may live in the 49 nations most vulnerable to all natural disasters and climate extremes by 2040, with Bangladesh, Democratic Republic of the Congo, Ethiopia, Kenya, Madagascar, Nepal, Nigeria, Pakistan, South Sudan, Sudan, and Uganda among the 11 most vulnerable to disaster-induced poverty.[44] While affluent nations with access to resources that can assist reduce environmental disasters frequently contribute the most to elements that raise the likelihood of environmental catastrophes, underdeveloped countries face the effects of such disasters more severely than their richer counterparts.[45]Capitalism's competitive dynamics encourage firms to constantly seek out new means of capital accumulation,

---

41 Richard M. Zoraster, "Vulnerable populations: Hurricane Katrina as a case study". *Prehospital and Disaster Medicine* 25 (2010): 74–78.

42 Ismail K. White, Tasha S. Philpot, Kristin Wylie, and Ernest McGowen, "Feeling the Pain of My People: Hurricane Katrina, Racial Inequality, and the Psyche of Black America". *Journal of Black Studies* 37 (2007): 523.

43 Rebecca Elliott, *Underwater: Loss, Flood Insurance, and the Moral Economy of Climate Change in the United States*. New York: Columbia University Press, 2021.

44 Andrew Shepherd, Tom Mitchell, Kirsty Lewis, Amanda Lenhardt, Lindsey Jones, Lucy Scott, Robert Muir-Wood. "The geography of poverty, disasters and climate extremes in 2030", 2013. ODI, accessed on 18 March 2022. Available online at: https://odi.org/en/publications/the-geography-of-poverty-disasters-and-climate-extremes-in-2030/

45 Étienne De Villemeur, Billette, and Justin Leroux, "Sharing the cost of global warming". *The Scandinavian Journal of Economics* 113 (2011): 758–783. See also Stephen M. Gardiner, "Ethics and global climate change". *Ethics* 114 (2004): 555–600.

which they do by extending their operations and investing in labor-saving technology. These attempts reduce labor costs but increase energy and chemical usage. However, economic progress and technological improvement are environmentally harmful since they increase pollution and resource exploitation.[46] Neoliberal reconstruction after natural disasters is rightly named as "wreak construction" to highlight the failures of market driven solutions targeting the poor and the disadvantaged groups through displacement and land grabbing.[47] The economy of every nation on earth has already been damaged by global warming. It has affected certain areas more severely than others, mostly in the tropics. The disparity between the world's richest and poorest nations is around 25 percent greater than it would be in a world without global warming.[48] During times of natural disaster, most individuals will leave their homes or workplaces to stay with friends and family. Those who seek refuge in public shelters are disproportionately members of underprivileged or minority groups.[49]

Due to ideological divides and national and international governmental ineptitude in addressing the problem, the negative effects of global warming on our planet are misrepresented and downgraded. From the media to the highest courts, the public is swayed by the inaction or denial of political and legal authorities who either diminish the severity of climate change's impacts or refuse to order effective responses to the problem. For instance, the climate change committee of the government determined that the Boris Johnson administration had dismally failed to meet its climate targets.[50] More vehemently, the supreme court in the US ruled, in favor of the state of West Virginia, that the Environmental Protection Agency does not have the authority to control carbon dioxide emissions from power plants.[51] This ruling from a higher court dampens our optimism and, while having an enor-

---

46 Liam Downey and Susan Strife, "Inequality, democracy, and the environment". *Organization & Environment* 23 (2010): 155–188.

47 Engin Atasay and Garrett Delavan, "Monumentalizing Disaster and Wreak construction: A Case Study of Haiti to Rethink the Privatization of Public education". *Journal of Education Policy* 27 (2012): 529–553.

48 Noah S. Diffenbaugh and Marshall Burke, "Global warming has increased global economic inequality". *Proceedings of the National Academy of Sciences* 116 (2019): 9808–9813.

49 Gillis Walter Peacock, Betty Hearn Morrow, and Hugh Gladwin (eds.), *Hurricane Andrew: Ethnicity, Gender and the Sociology of Disasters.* London: Routledge, 1997.

50 Michael Holder, "'We are worried': UK climate advisors slam 'shocking' lack of net zero delivery". *Business Green*, accessed on 5 July 2022. Available online at: https://www.businessgreen.com/news/4051995/worried-uk-climate-advisors-slam-shocking-lack-net-zero-delivery

51 "US supreme court rules against EPA and hobbles government power to limit harmful emissions". *The Guardian*, accessed on 4 July 2022. Available online at: https://www.theguardian.com/us-news/2022/jun/30/us-supreme-court-ruling-restricts-federal-power-greenhouse-gas-emissions

mous effect on the domestic front, has global implications that encourage opponents of climate change theory to take similar actions around the world.

Climate change, nuclear weapons, the high-tech arms race, wars, and our inability to save the earth bring us closer to future hazards that threaten the destiny of humanity. The question that must be answered is how we can strengthen our ability to withstand the effects of potential disasters. In this regard, social protests may be more effective than other means of garnering international attention. Local environmental social protests and their dissemination through the media around the world may raise global awareness of environmental issues, with international forces particularly influencing less democratic nations.[52] For those studying how to bounce back from devastating events, resilience is the overarching concept. A number of studies have found that the most significant change is an altered perspective on life, such as a diminished desire for material possessions and a heightened appreciation for family and personal resilience in the face of future tragedy.[53] However, these perspectives only provide an escape from the problem's origins and a short-term strategy for survival. This is why we wish to demonstrate in this book why human-environment relationship and the risks arising from this relationship is so crucial for addressing global insecurity. Different chapters in this book aim to deliver the message that humans are primarily responsible for triggering natural disasters, escalating the existential threat to humanity, and mismanaging post-disaster events, which all become repetitive in varying forms over time.

Human and Environmental Insecurities: A Multidisciplinary and Global Approach

The eight chapters in this book examine human and environmental insecurities in which disasters, various forms of violence, catastrophes, and environmental risks have plagued our planet on multiple levels. In daily life across different regions, environmental injustice has manifested itself similarly, albeit with different rhythms, and the most vulnerable communities have been those at the bottom of the social stratum. The contributors of this book come from a variety of academic fields, including history, political science, critical theory, sociology, and the arts.

---

52 Erin M. Evans, Evan Schofer, and Ann Hironaka, "Globally Visible Environmental Protest: A Cross-national Analysis, 1970–2010". *Sociological Perspectives* 63 (2020): 786–808.
53 Thomas E. Drabek, "Sociology of Disaster", in Kathleen Odell Korgen (ed.), *The Cambridge Handbook of Sociology, Volume 2: Specialty and Interdisciplinary Studies*. Cambridge: Cambridge University Press, 143.

Our objective is to expand our inter-disciplinary understanding of environmental injustice global insecurity by focusing on the disasters and hazards that threaten our lives and the future of humanity.

In their article, Henry Fowler, Cynthia Boyer, Kelly Tzoumis, and Kaitlyn Mansoorieh discuss nuclear fission experiments conducted by US scientists during World War II and their severe human and environmental repercussions, which have left a legacy of risk and policy issues unlike any other due to the enduring threat nuclear materials pose to human existence and their impact on future generations. Their article, entitled: "The Catastrophic Legacy of Nuclear Arming on Health: How Public Policy Failure in the US Shaped Generational Disaster", they argue that what has been largely ignored following the use of nuclear fission for both military and civilian purposes is twofold: (i) the impacts of extracting the uranium needed for weapons; and (ii) the ongoing disposal of nuclear waste from weapons and energy generation. They meticulously analyze the health effects of these wastes' legacy. In particular, their article examines the legacy left to the Navajo Nation, which endured adverse health effects and land contamination as a result of the US's extraction of uranium for the war effort using unreclaimed mines. In addition, the disposal of high-level nuclear fuel and waste is an unfinished public policy in the United States due to failures in locating a repository for these thousands of years-lasting materials. Using a case study approach for extraction and disposal, they establish a link between the production of nuclear weapons and the severe negative effects on human health and the environment. As many countries reexamine nuclear energy as a potential source of climate change, this chapter demonstrates what is frequently overlooked regarding the legacy of nuclear materials.

Rosalia Gambina's article, entitled "Sustainable Waste? Waste Colonialism and the 'Sustainable' Imaginary" examines "waste colonialism" as a destructive force and argues that it is supported and perpetuated by the "sustainable development" rationale. She aims to examine the corporeal dispossession that colonialism frequently entails – that is, the gradual seizure of a population's self-determining capacity over its health outcomes, which can and often does result in physical impairment. Gambina demonstrates conclusively that if populations in the Global North do not reduce their consumption and, consequently, waste production, an increasing number of populations in the Global South will be forced to consume waste to the detriment of their biological capacities. The chapter by Gambina discusses Rob Nixon's idea of "slow violence," which refers to violence that is incremental and accumulative, with its disastrous ramifications playing out over a variety of time scales. She focuses on electronic waste (also known as e-waste) and examines the global flows of e-waste to comprehend how waste colonialism operates in practice. In accordance with efforts to mitigate anthropogenic climate change, the sustainable development project aims to promote "clean" energy sources, from which we will inevitably generate more and more e-waste (e.g. lithium-ion batteries to fuel electric vehicles and store energy absorbed by solar panels). Consequently, the chapter points out that it is essential to pay close attention to the pattern that has emerged in waste disposal and, by extension, e-waste disposal. Gambina contends that the so-called developed world will continue to perpetuate a violent colonial power as the demand for sustainable development develops.

Svetlana Bokeriya explores the impact of global warming on the natural landscape and how it has influenced geopolitical shifts that threaten the Sahel region's security. In her article, "Security Challenges of the Climate Change in the Sahel Region", Bokeriya claims that increased demand for the country's natural resources reduces the country's ability to effectively manage the challenges and increases the likelihood of armed conflict. In comparison to other potential threats to global stability, the impact of climate change on international security can be forecasted with a relatively high degree of accuracy. Temperature changes have had a negative impact on Africans' health, means of subsistence, food industry, water availability, and safety.[54] The number of natural disasters such as floods and droughts has more than doubled in the last 25 years, resulting in more drought deaths in Africa than in any other region.[55] Bokeriya delves deeply into the Sahel region's climate change and security issues. In doing so, she identifies key international approaches to addressing Sahel climate and security issues. The analysis shows that the evolution of the UN and AU approaches reflects the security and environmental aspects that must be refined in order to reduce Sahel conflicts and, ultimately, have a positive impact on resolving the migration crisis, the terrorist threat, and reducing the total number of international conflicts.

Tatiana Konrad investigates climate change denial and the assault on science as forms of violence utilized by populists throughout the world. Konrad's chapter, "Violence in the Age of Environmental Crisis: Climate Change Denialism, the War on Science, Eco-Anxiety, and *Don't Look Up*," builds on Rob Nixon's significant idea of slow violence and highlights the need to re-envision the severe impact of violence today. In the era of environmental crisis, this reevaluation takes on added significance as Konrad examines the danger of invisible, non-immediate forms of violence that can have catastrophic results. Climate change skepticism and the war on science have polarized the global population and significantly impeded the collective action required to save the planet. Utilizing the black comedy *Don't*

---

54 Torsten Weber et al., "Analyzing regional climate change in Africa in a 1.5, 2, and 3 C global warming world". *Earth's Future* 6 (2018): 643–655.
55 Michael Berlemann and Max Friedrich Steinhardt, "Climate change, natural disasters, and migration – a survey of the empirical evidence". *CESifo Economic Studies* 63 (2017): 353–385.

Look Up (2021) to illustrate the dangers of climate change denial and the war on science, this chapter brings critical scholarship to the forefront. It attempts to respond to the following questions: Why does humanity refuse to accept the possibility of the end of the world? Why are humans unable to abandon the concept of living in the world as we know it in favor of environmentally friendly living and coexistence? Why do some individuals continue to deny the existence of environmental problems despite ongoing environmental degradation? Moreover, how can such conduct be interpreted as a form of violence? The responses to these important questions clarify both the complexity of the issues with which we are currently confronted and how the increasing global risks cannot be separated from human biases and populism's dangers.

Stefania Paladini's article, "Unsustainable Wars? The Use of Weapons in Lower Earth Orbit," examines the utilization of weapons in low-Earth orbit. Her unique viewpoint sheds light on an often-overlooked issue: how superpowers pose significant risks to space pollution and increase unpredictable risks to Earth by exploiting the absence of a framework for the use of weapons in space. The recent Russian ASAT (anti-satellite) missile test of November 2021, which purportedly caused more than 1,500 trackable orbital debris and forced ISS personnel to take refuge, is only the most recent of a series of wargame tests conducted in LEO (Earth's Lower Orbit). Even if ASAT tests are not the only cause of the worrying increase of debris in Earth's orbit, each incidence exponentially exacerbates what is now an obvious threat to space activities even before it becomes a military security issue. Given the readiness of nations to test their own ASAT missiles, the increasing number of space actors makes it even more necessary to handle this situation (as the not-so-recent examples of China and India prove). The legal foundation for the employment of weapons in space specifically prohibits WMDs (Weapons of Mass Destruction) but not ASATs, according to Paladini. This chapter meticulously uncovers the danger of unrestricted use of unconventional weapons in space, as well as the relevance of ensuring that space remains the domain of humanity, as it was always intended to be used for the common good.

Ipshita Bhattacharya's chapter, titled "Arctic, the Zone of Geo-Politics: Risks of Living in the Anthropocene," examines risks associated with the Arctic region, which have become more critical as a result of climate change and develops critical perspectives on political conflict and competition to control the Arctic. The Arctic is undergoing a rapid systemic transformation with far-reaching economic, climatic, social, political, and security consequences that are poorly understood. In addition, the Arctic is becoming a political powder keg as the United States, Russia, and China seek to escalate their claims to the region's natural resources via strategic and military means. U.S. interest in the Arctic has been piqued by China's "Belt and Road Initiative" and its cooperation with Russia in establishing new maritime routes. This chapter examines in depth how China, the United States, and Russia's political and security engagements pose a threat to the region and what could occur if they fail to manage and resolve future Arctic challenges.

The climate catastrophe has a damaging effect on ecologies and bodies alike, one that is not limited to the material components of biological existence but affects the very nature of experience and its manifestation, which has special implications for aesthetic expression in relation to experience itself.[56] Art serves to manifest our fears, and its various forms can be a potent indicator of our emotions, anxieties, and future uncertainties. Ben Jack Nash is a sculptor and installation artist whose artwork and performance work with high-resolution images in his chapter demonstrate an artist's preoccupation with and artistic manifestation of catastrophe. In his "What is the Matter with Catastrophes" reflective essay, Nash provides a series of photographs derived from a performance-based art installation which he delivered during our conference. A central component of the work is a time-lapse film depicting an embryonic chicken in its early stages of development. The film begins with the yolk, an abstract, sun-like, monochrome yellow image that commences to mutate and expand. From a network of red capillaries to a beating heart, we observe the embryo develop into a fully developed fetus with eyes and a beak. Uncertain of the fate of the fetus, the film abruptly returns to the beginning. Melting ice, land erosion, species extinction, bleached coral, forest fires, and desertification represent physical matter in its more solid and functional form of uncertainty dramatically shifting towards more abstract states and much more quickly than normal evolution rates. The images in Nash's essay convey aesthetically that the climate change we are currently experiencing is the result of human activity, which has placed our future in grave danger and uncertainty.

These difficult times necessitate a new way of thinking regarding the effects of climate change on the legal system and the humanities. The change with climate change is substantive, in that we must adapt the law's content and respond to current environmental challenges. The chapter of Matteo Nicolini, entitled "Law and Humanities in a Time of Climate Change" rejects the current positivistic approach to climate change; going beyond the boundaries of legal studies, it conducts cross-disciplinary research across law and the humanities. This article offers fresh insight by proposing that the fields of law and the humanities may be useful in assessing how climate change affects the legal spectrum and the nonfiction genre by providing new opportunities and updated perspectives. This chapter focuses on the legal framework and the recently developed nonfiction literary genre of "climate-

---

56 Michael Richardson, "Climate Trauma, or the Affects of the Catastrophe to Come". *Environmental Humanities* 10.1 (2018): 1.

change pop-science." Nicoloni's analysis of the legal framework, alongside essays, pamphlets, and other writings, sends a powerful message to the public that it is time to act in response to the climate and ecological emergency.

Each of the eight chapters examines a different topic, and the authors' approaches vary due to their diverse academic backgrounds. Nevertheless, all authors agree on the principle that environmental injustice and its catastrophes pose a threat to the future of humanity and that the current approach to addressing this global issue lacks serious structural capacity. The authors' shared argument and the outcomes of each chapter indicate two major conclusions. First, environmental injustice and its associated catastrophes are no longer dismissible by national and international authorities. The most immediate evidence for this is the historical lessons and the current dire circumstances. Second, the absence of structural capacity may necessitate an increase in social mobilization in order to encourage more conservative politicians to stop ignoring the urgency of environmental injustice and its catastrophes. This book connects various disciplines through the use of bibliographic notes, historical case studies, personal reflections, and original research. In doing so, we emphasize the two key messages through which we seek the possibility of a better future for humanity.

Henry Fowler, Cynthia Boyer, Kelly Tzoumis, and Kaitlyn Mansoorieh

Chapter I
The Catastrophic Legacy of Nuclear Arming on Health: How Public Policy Failure in the US Shaped Generational Disaster

Introduction

This case study examines the impacts and consequences from the weapons of mass destruction that are often ignored or underestimated by US policymakers and climate change activists. It also challenges human society to question the overlooked impacts to the current energy race for achieving carbon neutrality. Some lessons learned from the consequences of developing a nuclear weapons arsenal can be instructive into the future as the world seeks alternatives to fossil fuels for energy. The research begins with a historical context on the use of weapons production and development. This is necessary for understanding the legacy of nuclear catastrophe from the mining of uranium ore and disposal of nuclear wastes. Much of this research is not well-articulated and is hastily dismissed in the public discussion on how to move away from fossil fuels, particularly by the advocates focused on addressing climate change. Specifically, the costs and impacts of the nuclear legacy to the Navajo Nation (Diné is preferred name of the people) is examined because it is instructive for future energy choices taking place now. This research challenges the notion of nuclear energy as solving the climate change problem because it is zero net carbon energy source. This research concludes that this reframing of nuclear energy to address climate change is too narrow of a policy approach. The nuclear industry has been effective at recasting nuclear energy as an environmentally-friendly alternative to fossil fuels in what has been a sort of renaissance of the technology by "green-washing." The waste produced during extraction and generation is questioned due to looking at only the climate change impacts rather than a holistic human health impact lens which lacks the waste disposal, and ultimately, its sustainability.

Historical Context

During World War 2, scientists pursued a weapon of mass destruction relying on a new technique in basic physics never tested before. It focused on the splitting the

https://doi.org/10.1515/9783111081687-002

nucleus of a Uranium atom. Uranium ore is a naturally occurring radioactive element found at elevated levels across the Diné lands and the entire Colorado Plateau. While the equations for energy produced from Einstein's $E=mc^2$ theoretical model had existed at least since its publication in 1905, the use of the nuclear weapon demonstrated it viability. It was the birth of nuclear fission in places like Los Alamos, NM which led to the creation of a nuclear weapon of mass destruction that had the power of no previous weapon produced. This weapon has been casted as a tool that ended the war because the allies of Europe and the US were suffering high causalities, and some military scholars continue to conclude these countries were at risk of losing the war without the launch of a weapon of mass destruction.

The first, and only time used in history, nuclear weapons of mass destruction were launched on Hiroshima and Nagasaki, Japan in August 1945 by the US. The development and testing of these weapons were accomplished by academic scientists in conjunction with the support of the military in the US under great secrecy. Then, shortly after the use of nuclear weapons in 1945 by the US, a global nuclear arms-race began. History has labelled this the "cold war period" that was dominated by a diplomacy of isolation and nationalism. The result of this failed diplomacy culminated in the overproduction of these weapons of mass destruction that remain today. Across the world, the development of a stockpile of atomic weapons started rapidly along with a proliferation of these weapons worldwide. Attempts to reduce nuclear arms took place in 1968 with the Nuclear Nonproliferation Treaty (NPT) and the Comprehensive Nuclear Test Ban Treaty in 1996. Prior to the NPT, nuclear warheads totaled to tens-of-thousands for the combined arsenal of the US and the former Soviet Union. However, while arms reduction was effective under these treaties, there is still significant expansion of countries pursuing these weapons. For instance, Iraq initiated a nuclear program under former leader Saddam Hussein before the 1991 Persian Gulf War. North Korea withdrew from the NPT in 2003 and continues to test nuclear devices. Iran and Libya continue to develop nuclear devices in violation of the treaty, and Syria is also suspected of doing the same. The current estimate of global nuclear warhead inventory in 2021 includes 6,257 for Russia, and 5,550 for the US (see Chart 1). While the US and Russia have warheads in the thousands, nuclear weapons have spread to eight other countries with a warhead stockpile in the hundred or less (China, UK, France, India, Pakistan, India, Israel, and North Korea). Today, the world's nuclear-armed states possess a combined total of nearly 13,080 nuclear warheads. More than 90% are

```
6257
5550




                           350     290     225     165     156      90    40–50

Russia  United   China  France  United  Pakistan India  Israel North Korea
        States                  Kingdom
```

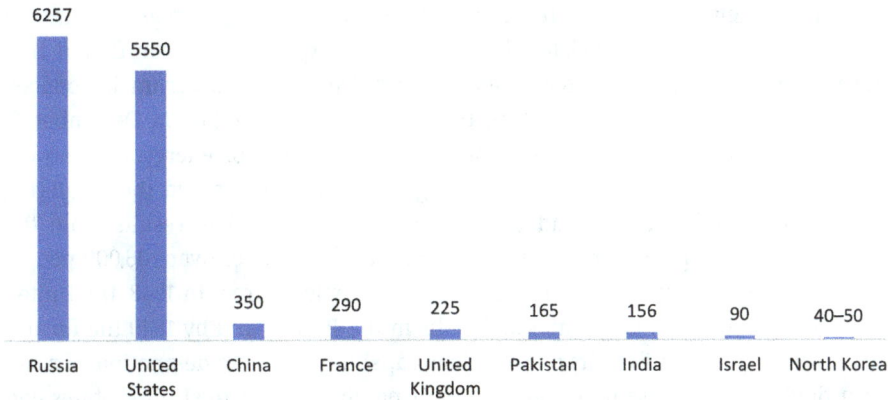

Figure 1: 2021 Estimated Global Nuclear Warhead Inventories.
Source: "Nuclear Weapons: Who Has What at a Glance," Arms Control Association, last reviewed January, 2022, https://www.armscontrol.org/factsheets/Nuclearweaponswhohaswhat

owned by Russia and the US. Approximately 9600 warheads are in military service with the rest awaiting dismantlement.[1]

A new foreign policy called *Mutually Assured Destruction* (ironically referred to as MAD) was born out of fear of annihilation from the use of this weapon. Over time countries experienced the economic burden of this arms-race and the international security problems of a nationalistic foreign policy. Some nation-states slowly stepped away from isolationism. The US recognized the imperative of international negotiation to limit future production and reduce the arms inventory. This included a series of de-escalation of nuclear weapons through compromises such as the Strategic Arms Limitation Talks (SALT) I and II during the 1970s, and more recently a variety of nuclear weapon agreements among nations. The number of nuclear warhead inventories have reduced significantly since the build up from 1945 to the mid-1980s. By the end of the 1980s–2021, the nuclear war head inventory has decreased about 80%.[2] Today, from a series of bilateral arms control agreements, the US deploys 1,357 and Russia deploys 1,456 strategic warheads on bombers and missiles from their larger inventory. The most recent agreement caps each country at 1,550 strategically deployed warheads.[3]

---

1 Hans M. Kristensen and Shannon N. Kile, Estimated Global Nuclear Warhead Inventories. Stockholm International Peace Research Institute, and the U.S. Department of State, 2021.
2 Hans M. Kristensen, Robert S. Norris, and Matt Korda, Estimated Global Nuclear Warhead Inventories 1945–2021. Federation of American Scientists, 2021.
3 Hans M. Kristensen, Shannon N. Kile, Estimated Global Nuclear Warhead Inventories. Stockholm International Peace Research Institute, and the U.S. Department of State, 2021.

Immediately after WW2, there was an *Atoms for Peace Program* initiated under former President Dwight Eisenhower that expanded nuclear fission into the purposes of generation for civilian energy. This was a program introduced by the president in a speech before the UN General Assembly on December 8, 1953, where nuclear fission was re-casted to a civilian use for energy. Eisenhower was concerned about nuclear fission being used for weapons and the escalating arms race between the US and the former Soviet Union. The results from the WW2 nuclear weapons were catastrophic with an estimate of over 106,000 people mortally impacted and at least another 110,000 people injured. In 1948, the United States had tested even larger atomic bombs in the Pacific, and by 1949 the former Soviet Union had achieved its own nuclear capability with the detonation of a nuclear device. In response to the Soviet atomic bomb program, the United States had embarked upon a program to develop an even more powerful weapon, the hydrogen bomb, which promised explosive power in the range of millions of tons of traditional explosives. The United States successfully detonated a hydrogen weapon in November 1952; just a few days before Eisenhower won the presidency. The 10-megaton blast had destroyed the test island of Elugelab, creating an underwater crater 1,500 yards in diameter.[4]

This was a threefold impact for the Diné. The first impact was this drive for victory of war in testing and using nuclear weapons. The second impact was the arms-race after the war. The third impact was the implementation of civilian nuclear energy for utility power that occurred after the war. Several consequences to both human and environmental health have resulted which cast a legacy of risk and policy issues like no other energy source. The reason for this legacy is linked to the long-lasting risk of nuclear materials in the form of isotopes that can dramatically impact human health as well as be passed to future generations. This waste crosses the cell walls of life forms to denature the nucleic chromosomes of human existence, and can remain an environmental pollutant for thousands, if not tens-of-thousands of years depending on its half-life of the isotope.

What has been mostly ignored after the use of nuclear fission by both military and civilian uses is: 1) the impacts of extracting the uranium needed for the weapons, and 2) the lack of disposal for nuclear wastes from weapons and energy generation that continues today. This paper focuses on the legacy of human health impacts from these wastes. Specifically, this research will review the legacy to the people of the Diné (Navajo Nation) who suffered the health impacts and land contamination from un-reclaimed mines by the US from the extraction of uranium.

---

4 "Atoms for Peace". *Dwight D. Eisenhower Presidential Library*, accessed February, 2022. Available online at: https://www.eisenhowerlibrary.gov/research/online-documents/atoms-peace

Also, the civilian energy disposal of high-level nuclear fuel and wastes is an unfinished public policy in the US due to failures of locating a repository for these long-lasting materials. The costs to human health and the environment are explored as hidden consequences and often ignored impacts tied to both the beginning of the extraction process and the disposal after use of the materials. This research provides insights from the past that can assist with ongoing and future decisions about the use of nuclear fission for war and civilian energy production. Using a case study approach on extraction and disposal, important findings are exposed that are often unconnected to both the production of nuclear weapons and energy generation. This disconnect continues today in the nuclear remediation policies of the US. These results from this case study analysis also provide what is often overlooked as the world is reconsidering nuclear energy as a potential source for climate change.

A Case Study: The Nuclear Legacy of Uranium to the Diné People

Beginning in the 1940s and continuing for approximately 40 years, private companies extracted uranium ore. Approximately 30 million tons of uranium ore were extracted from mines on the Diné lands surrounding the Four Corners area of the US from 1944 to 1989.[5] These lands include over 27,000 square miles spanning the states of Arizona, New Mexico, Utah. Of the 4–6 sources of uranium in the world, this area is one of the most enriched locations. Until 1970, the US purchased most of the uranium ore from Diné lands to meet the demand for developing the nation's first atomic bomb and the subsequent production of weapons for the US nuclear weapons stockpile. Uranium production directly affected the Diné people who worked in the mines to extract the uranium ore or who lived near mines. Active uranium mining on the reservation ceased by 1986, and companies often abandoned the mines. Nearly 30 years later, the Diné people continue to live with the environmental and health effects from mining operations. Lung cancer, bone cancer, and impaired kidney function can result from exposure to elevated levels of uranium and other radionuclides associated with the mining and processing of the ore. Today, the US Environmental Protection Agency (EPA), the lead agency

---

5 United States Government Accountability Office, *Uranium Contamination: Overall Scope, Time Frame, and Cost Information is Needed for Contamination Cleanup on the Navajo Reservation.* GAO-14–323, May 2014: U.S. Environmental Protection Agency, *Ten Year Plan: Federal Actions to Address Impacts of Uranium Contamination on the Navajo Nation 2020–2029.* 2021.

for cleaning up wastes in the US, have identified 523 abandoned uranium mines on or near the Navajo Nation lands. More than 1,000 features such as mine openings and waste piles have also been assessed. There are 485 abandon uranium mines on the Nation lands, and 38 mines within 1 mile of the nation in the states of New Mexico (230 mines), Arizona (6 mines) and Utah (2 mines).[6]

Since WW2, the Diné people have lived with the environmental and health effects of uranium contamination from this mining. In 2008 and again in 2014, federal agencies[7] adopted both Five-Year Plans encouraged by Congress after testimony from the Navajo Nation in 2007 before the US House Committee on Oversight and Government Reforms on the impacts of both mining and milling the materials on the Navajo people.[8] It was this hearing that spurred the first Five-Year Plan starting in 2008.[9] The US Government Accountability Office (GAO), a research investigatory arm of the US Congress, reviewed the first Five-Year Plan. It found that an estimate of the overall scope of the remaining work, time frames, and costs to fully address the uranium contamination cleanup was needed. [10] The agencies along with the Diné organizations prepared a second Five-Year Plan in 2014, which led to the more recent Ten-Year Plan active until 2029.[11]

These plans identified targets for addressing contaminated abandoned mines, structures, water sources, former processing sites, and other sites on the Diné lands. Federal agencies also provide funding to Navajo Nation agencies to assist with the remediation and recovery work. In 2020, these agencies along with the Indian Health Service issued the Ten-Year Plan titled *"Federal Action to Address Impacts of Uranium on the Navajo Nation."*[12] The US federal government was the sole consumer and customer of this uranium ore until 1970, with sales to commercial

---

6 U.S. Environmental Protection Agency, *Ten Year Plan: Federal Actions to Address Impacts of Uranium Contamination on the Navajo Nation 2020–2029.*

7 The agencies include the US Environmental Protection Agency, the US Department of Energy, the Agency for Toxic Substances and Disease Registry, the Nuclear Regulatory Commission, and the US Department of Interior Bureau of Indian Affairs. These agencies were joint by the Indian Health Service.

8 *The Health and Environmental Impacts of Uranium Contamination in the Navajo Nation, Before the Committee on Oversight and Government Reform*, 110th Congress. 2007.

9 U.S. Environmental Protection Agency, *Five Year Plan: Federal Actions to Address Impacts of Uranium Contamination on the Navajo Nation 2014–2018.* 2014.

10 United States Government Accountability Office, *Uranium Contamination: Overall Scope, Time Frame, and Cost Information is Needed for Contamination Cleanup on the Navajo Reservation.* GAO-14-323, May 2014.

11 U.S Environmental Protection Agency, *Ten Year Plan: Federal Actions to Address Impacts of Uranium Contamination on the Navajo Nation 2020–2029.*

12 U.S. Environmental Protection Agency, *Ten Year Plan: Federal Actions to Address Impacts of Uranium Contamination on the Navajo Nation 2020–2029.*

industry that began in 1966. Many of the Diné residents worked at the mines and lived nearby. By 2019, assessments were completed at 113 of these uranium mines, and it included 43 priority sites that were located near residential homes. EPA secured funding over $1.7 billion from a combination of enforcement agreement and settlements with the Navajo Nation from private companies. This funding ensures work on 230 of the 523 abandon sites and an additional 16 sites from mines plus the abandon mines called the Tronox Mines[13] which were from previous owner Kerr McGee Cooperation and its successor Tronox (a company now bankrupt). The Kerr-McGee Corporation was founded in 1929 as an energy company involved with oil and gas exploration and production, and uranium mining. The company left several abandoned uranium mine sites, including contaminated waste rock piles. EPA received almost $1 billion from litigation to address over 50 mines and the Navajo Nation received $40 million from the settlement.[14] From 2008–2019, the Navajos have received access to piped water to 3809 homes in the six abandon mining regions.[15] Over 1100 structures have been accessed for contamination and at least 50 structures remediated.

Once work ceased at a site, companies often abandoned the mines, leaving the waste rock piles in place without conducting any cleanup or posting signs warning about the dangers of contamination or physical hazards. The extracted ore was sent to an off-site processing facility called a mill. At the mill, the mined uranium ore was crushed, ground, and then fed to a leaching system that produced yellow slurry – called *yellowcake* – that was further processed for use in nuclear weapons or, as of the mid-1960s, for use in nuclear power plants. The leaching system left a waste product known as mill tailings that retained some toxic contaminants. The tailings were of a sandy consistency and mixtures of tailings and water were placed in unlined evaporation ponds at the mill site. The US Department of Energy (US DOE) estimates that millions of gallons of water contaminated by mill tailings were released into the groundwater over the life of the sites through the unlined

---

13 "Navajo Nation: Cleaning Up Abandoned Uranium Mines, Tronox Abandoned Uranium Mines", *U.S Environmental Protection Agency*, last modified July 30, 2021. Available online at: https://www.epa.gov/navajo-nation-uranium-cleanup/tronox-abandoned-uranium-mines

14 "Tronox Abandoned Uranium Mines," *United States Environmental Protection Agency*, accessed February, 2022. Available online at: https://www.epa.gov/navajo-nation-uranium-cleanup/tronox-abandoned-uranium-mines. "$2 Billion Funds Headed for Cleanups in Nevada and on the Navajo Nation from Historic Anadarko Settlement with U.S. EPA," *United States Environmental Protection Agency*, last modified November 29, 2017. Available online at: https://archive.epa.gov/epa/news releases/2-billion-funds-headed-cleanups-nevada-and-navajo-nation-historic-anadarko-settlement. html

15 U.S. Environmental Protection Agency, *Ten Year Plan: Federal Actions to Address Impacts of Uranium Contamination on the Navajo Nation 2020–2029.*

ponds. In 1978, the US passed legislation to control the mill tailings. [16] In addition, on July 16, 1979, the largest release of radioactive materials in the United States occurred when a dam on one of the evaporation ponds broke at a processing site near Church Rock, New Mexico, resulting in the release of 94 million gallons of radioactive waste to the Puerco River, which flowed through nearby communities.[17] The US DOE is responsible for the long-term maintenance of four uranium-ore processing milling sites on the Diné lands. The Nuclear Regulatory Commission (NRC) has regulatory authority over these former mill and processing sites that are managed by the US DOE. These mill and mining wastes were used by some Navajo residents for building materials of their homes.[18] Also, dust from these wastes is suspected to have contaminated the clothing and personal belongs brought into the residential homes. One of the most important policies for the Diné people has been the Radiation Exposure Compensation Act of 1990 (RECA).[19] This law provided compensation to people impacted by the uranium. In 2000, the law was expanded.

Addressing Health Impacts: The Radiation Exposure and Compensation Act (RECA)

In 1990, the US passed legislation titled the *Radiation Exposure and Compensation Act* (RECA).[20] This law compensates those impacted uranium including the Navajo uranium miners who suffered illness from this work while the US was the sole purchaser of uranium. The genesis of the legislation was based on the high rates of illnesses among miners, workers at the Nevada Test Site (NTS) where testing was performed on weapons, and communities living downwind from the NTS. This law was amended in 2000 based on complaints from those applying for compensation like the uranium miners who failed to be compensated because the standard dose exposure was set very high at six to 15 times the elevated risk and discounted miners who smoked.[21] The first claims were processed in 1992.

---

16 Uranium Mill Tailings Radiation Control Act of 1978, Pub. L. No. 116–260, (2020).
17 United States Government Accountability Office, *Uranium Contamination: Overall Scope, Time Frame, and Cost Information is Needed for Contamination Cleanup on the Navajo Reservation*.
18 U.S. Environmental Protection Agency, *Ten Year Plan: Federal Actions to Address Impacts of Uranium Contamination on the Navajo Nation 2020–2029*.
19 Radiation Exposure Compensation Act of 1990, 1990, https://www.justice.gov/civil/common/reca.
20 Radiation Exposure Compensation Act of 1990, 1990.
21 Doug Brugge and Rob Goble, "The Radiation Exposure Compensation Act: What Is Fair?" *New solutions* 13, no. 4 (2004): 385–397.

RECA was a revolutionary piece of legislation when passed in 1990 because it recognized the responsibility for the harm done to those exposed to uranium and, more importantly, its purpose was stated – to make partial restitution which also included an apology by the US. This is the most profound example of reparations in US history that is something mostly denied to other overburdened groups like those descendants of slavery in the US or colonialism. However, the 1990 RECA program contained many operational flaws. Moreover, a report by the Advisory Committee on Human Radiation Experience in 1996 found that the US Public Health Service violated research ethics by not informing uranium mining workers who enrolled in the study of their health risks.[22]

By the end of 2000, the RECA program received 7,819 applications for compensation.[23] The processing of the claims was dismal. There was only the acceptance of 46% and rejection of 46% applications, with 8% pending at the time of the investigation.[24] The primary reason for denials were that the uranium miners and other claimants like the on-site participant or downwinder was that they did not meet the minimum exposure to radiation requirements or did not contract a disease that was eligible for compensation. After reviewing RECA legislation, the regulations, and complaints by the uranium miners to the 1990 requirements, researchers found that there were significant deficiencies in the 1990 legislation with only some being corrected in the 2000 amendments.[25] Immediately after passage of the 2000 amendment, it became clear that the US Congress neglected to appropriate funding to pay the claims. RECA claims were most frequently denied because the disease contracted by the victim is not specifically designated as eligible for compensation under the RECA program.[26] Treasury paid $245.1 million in compensation to 5,150 victims or their survivors, but later awards were not paid by the end of fiscal year 2000 because the money in the Trust Fund was depleted. This led to a spectacle of the US Department of Justice sending out notices the equivalent of an "I-owe-you" resulting in checks being delayed until July 2001.[27] The Bush Administration announced it wanted to stop the payments to mill workers and ore haulers

22 United States, *Final Report of the Advisory Committee on Human Radiation Experiments.* New York: Oxford University Press, 1996.
23 United States Government Accountability Office, *Radiation Exposure Compensation: Analysis of Justice's Program Administration.* GAO-01–1043, Sep. 2001.
24 United States Government Accountability Office, *Radiation Exposure Compensation: Analysis of Justice's Program Administration.*
25 Brugge and Goble, "Radiation", 386.
26 United States Government Accountability Office, *Radiation Exposure Compensation: Analysis of Justice's Program Administration.*
27 Brugge and Goble, "Radiation", 394.

until more scientific research was completed. However, by 2002, these workers did receive compensation. Basically, the researchers concluded that RECA was flawed because it simultaneously apologized, set a highly stringent criteria, and then placed the burden of proof on the victims in the manner repeating the policy errors when the 1990 RECA was first passed.[28]

In 2019, the US Department of Justice reported that since the RECA Program commenced operations in April 1992, 48,049 claims have been filed and over $2.28 billion has been awarded in connection with 34,942 approved claims (through September 30, 2018). Compensation has been awarded to individuals residing in all 50 states, as well as several foreign countries. Residents of the "Four Corners" region of the US southwest region filed the majority of the claims and received awards valued at over $1.6 billion during the life of the program. Of the more than $2 billion in awards, approximately $310 million has been awarded to Native American claimants and distributed among members of 20 different tribes, and approximately $330 million has been awarded to veterans, civil servants, and contractors who participated onsite in atmospheric nuclear tests.[29] These claims focus on the extraction from mining of the materials on Diné lands. The recent information on claims from 2022 by the Department of Justice show that the on-site participant (58%), uranium miner (62.9%) and miller (75.1%) have lower acceptance rates than the others expect for the ore transporter and downwinder. None of the rates of acceptance come close to 95% or higher (see Table 1 below).[30]

In addition to extraction and processing of the uranium as a source of risk to human health, the next section turns to the other end of the nuclear fission cycle – waste disposal.

---

28 Brugge and Goble, "Radiation", 395.
29 The United States Department of Justice, *Radiation Exposure Compensation Act Trust Fund: FY 2020 Budget & Performance Plan*, 2019. U.S. Congressional Research Service, The Radiation Exposure Compensation Act (RECA): Compensation Related to Exposure to Radiation from Atomic Weapons Testing and Uranium Mining. R43956, January 13, 2021.
30 "Radiation Exposure Compensation System: Awards to Date 3/1/2022", United States Department of Justice, last modified March 1, 2022. Available online at:https://www.justice.gov/civil/awards-date-03012022

Table 1: RECA Summary of Claims by February 28, 2022.

Radiation Exposure Compensation System
Summary of Claims Received by 02/28/2022

Claim Type	Pending	Approved	% Approved/ of Disposed	$ Approved	Denied	Total
Downwinder	205	24,771	83.6	$1,238,520,000	4,861	29,837
Onsite Participant	68	5,168	58.0	$377,321,401	3,735	8,971
Uranium Miner	105	6,796	62.9	$678,874,560	4,004	10,905
Uranium Miller	38	1,881	75.1	$188,100,000	623	2,542
Ore Transporter	6	389	70.3	$38,900,000	164	559
Total:	422	39,005	74.4	$2,521,715,961	13,387	52,814

Updated March 1, 2022

Source: "Radiation Exposure Compensation System: Claims to Date Summary of Claims Received by 3/1/22) All Claims," Department of Justice Civil Division, last reviewed March, 2022, https://www.justice.gov/civil/awards-date-03312022

The Issue of High-Level Nuclear Waste Disposal Remains

Several advocates from the nuclear industry and the US Department of Energy as well as financier Bill Gates have proposed nuclear energy as the high generating energy source to replace fossil fuels.[31] In fact, some view nuclear energy as a necessary component to the global energy mix of options to decarbonize and alter climate change patterns.[32] Advanced nuclear energy technologies are evolving and moving toward commercialization. These new technologies are smaller with some proposed to be modular and generate less waste. The US Department of Energy (DOE) is working toward prototypes of several nuclear reactors. One is an advanced small module reactor that they characterize as safe, clean, and affordable nuclear power. It is anticipated that this small module reactor design will be cer-

---

31 Catherine Clifford, "How Bill Gates' company TerraPower is building next-generation nuclear power". *CNBC* April 8, 2021. Available online at: https://www.cnbc.com/2021/04/08/bill-gates-terra power-is-building-next-generation-nuclear-power.html
32 Darren McCauley, Raphael J. Heffron, Hannes Stephan, and Kristen Jenkins, "Advancing Energy Justice: the Triumvirate of Tenets". *International Energy Law Review,* 2013.

tified by the US Nuclear Regulatory Commission, the federal agency that has the sole authority to certify nuclear designs in the US, from an application submitted in March 2017. The DOE is supporting the siting of the nation's first 12-module powerplants at their Idaho National Laboratory. Operation is expected to begin in 2029. However, the failure of being able to dispose of high-level nuclear wastes in the US is a case that yields several lessons that climate change activists could learn from when proposing nuclear energy as the solution to achieving a zero-carbon alternative to fossil fuels. In the US, the uncertainty for nuclear energy industry due to public opposition towards the energy source and its waste disposal, plus the large costs for building and retiring the power plants, makes its future uncertain in many countries, but certainly not all countries. The risk of public acceptance of these new reactor designs according to the US Energy Information Administration, an agency under the US DOE, view the future of nuclear energy in the US as uncertain.[33] With the recent political conflicts in Ukraine that have involved the former nuclear reactors at Chernobyl, this issue of risk has resurfaced in the media and perhaps the public regarding the costs to nuclear disposal.

In the past, the older designed nuclear powerplants produced a significant amount of energy with one facility. Post-WW2, nuclear energy was sought after as the goldmine for large scale energy production for civilian purposes. However, there are many nuclear powerplants that are aging and need decommissioning globally for their retirement that will generate significant wastes of several types. The International Atomic Energy Agency[34], an organization of the United Nations, states that "currently, about two-thirds of nuclear power reactors have been in operation for over 30 years, highlighting the need for significant new nuclear capacity to offset retirements in the long-term. Uncertainty remains regarding the replacement of the large number of reactors scheduled to be retired around 2030 and beyond, particularly in Northern America and Europe." However, nuclear power has recently been framed as a form of carbon neutral energy source that can be used to meet the needs for high energy consumption while having no impact on climate change. The International Energy Agency (IEA), a collation of 30 countries seeking to ensure reliable, affordable, and clean energy was founded in 1974 to help countries coordinate the disruptions to oil supply, works on issues of energy generation today. According to the IEA[35], the use of nuclear power has

---

33 "Homepage", *U.S. Energy Information Administration*, accessed February 2022, https://www.eia.gov/.

34 International Atomic Energy Agency, *Energy, Electricity and Nuclear Power Estimates for the Period up to 2050*, 1–40. Paris: IAEA, 2020. International Atomic Energy Agency. *Energy, Electricity and Nuclear Power Estimates for the Period up to 2050*, 3, 9.

35 International Energy Agency. *Nuclear Power in a Clean Energy System*. 2019.

avoided more than 60 gigatons of carbon dioxide emissions over the past 50 years. Nuclear energy has recently been reframed as an energy source to combat climate change.

Nuclear energy is one of the most uncertain energy sources in the mix of carbon neutral options. The estimates range widely on the global dependency on nuclear energy into the future. IAEA[36]claims that the range for global nuclear energy generation for electricity to 2050 can be anywhere between a high of 82% increase to a low of 7% decrease compared to 2019 global dependency of nuclear power. Part of the reason for this wide variation for the future estimate is the low price of natural gas and the fast growth of renewable energy technologies. Nuclear energy is one of the most expensive sources of energy to build as a generating facility with high upfront costs. In some countries like the US, construction times and regulatory approvals can be cost sinks for nuclear energy projects while China's and France's experience is much different.

IAEA[37] reports that at the end of 2019, globally there were 443 nuclear power reactors operational, with a total net installed power capacity of 392 gigawatts. In addition, 54 reactors with a total capacity of 57 gigawatts were under construction. Six new reactors were connected to an energy grid and 13 reactors retired. Construction began on five new reactors. Nuclear power accounted for 10.4% of global total electricity production in 2019, an increase of 0.2 percentage points from the previous year and the first increase since 2015. The US produces most of the world's nuclear electricity followed by France and China. However, France is 70% dependent of its total electricity production on nuclear energy, then the Ukraine and Slovakia. The US is only 19.3% dependent and that is expected to dramatically decline to 11% without the replacement of its aging nuclear power plants. Also, the OECD reported in 2020 that Germany plans to phase out nuclear power by the end of 2022, and there are potential reactor closures in France, Korea, Sweden, UK, and the US. In France, the energy transition toward green growth plans for a total of 14 power reactors to reduce its dependency on nuclear energy from 70% to 50% by 2035.[38]

It is non-OECD countries who are leading the way on building new nuclear reactors globally. China began construction on only one new nuclear reactor in 2019, the first since 2016, but in the next several years it is expected to launch several

---

36 International Atomic Energy Agency, *Energy, Electricity and Nuclear Power Estimates for the Period up to 2050*, 1–40.
37 International Atomic Energy Agency, *Energy, Electricity and Nuclear Power Estimates for the Period up to 2050*, 1–40. International Atomic Energy Agency, *Energy, Electricity and Nuclear Power Estimates for the Period up to 2050*, 3, 9.
38 International Energy Agency, *World Outlook 2020*. OECD Publishing, Paris, 2020.

new projects to reach a total capacity of as much as 110 gigawatts by 2030.[39] In India, new reactors have been limited during the last few years, but the country aims to construct 21 new nuclear power plants by 2030.

The failure in the US to site a high-level nuclear waste repository is one the most costly and well-known public policy failures in history. In 2008, the application for a national deep geological burial site called Yucca Mountain located in the state of Nevada, was sent for approval to the Nuclear Regulatory Commission. Note, the state of Nevada generates zero nuclear energy, and does not have any significant crude oil, natural gas, or coal reserves. It is undergoing a massive growth of solar photovoltaic energy at the rooftop and ground levels for electric utilities. Most of Nevada's energy comes from natural gas which accounts for almost two-thirds of its electricity in addition to renewables that include geothermal and hydroelectric.[40]

In 2009, the Obama Administration terminated the Yucca Mountain project and formed a commission to recommend other options and processes for high-level nuclear waste disposal. A report from the Blue Ribbon Commission on American Nuclear Future[41] formed by President Obama, clearly stated that the US's nuclear waste management program was at an impasse. It basically took a process for siting high-level nuclear waste which began discussion in the 1950s with the National Academy of Sciences and then identified Yucca Mountain in 1987 under the Nuclear Waste Policy Act[42], into a start over process.

Today, nuclear powerplants in the US do not have a repository to dispose of the waste after decades of scientific inquiry and public policy programming. According to the *Blue Ribbon Commission Report*,[43] nuclear utilities are assessed a fee on every kilowatt-hour of nuclear-generated electricity as a quid pro quo payment in exchange for the federal government's contractual commitment that was to begin accepting commercial spent fuel by January 31, 1998. These fee revenues go to the government's Nuclear Waste Fund, which was established for the purpose of covering the cost of disposing of civilian nuclear waste and ensuring that the waste program would not have to compete with other funding priorities. The an-

---

39 International Energy Agency, *Global Energy Review 2021*. Paris, 2021.
40 Sarah Zhang, "Nevada Fights the Latest Attempt to Give It the Nation's Nuclear Waste". *The Atlantic*, April 26, 2017. Available online at: https://www.theatlantic.com/science/archive/2017/04/nuclear-waste-yucca-mountain-hearing-states/524418/
41 Blue Ribbon Commission, "Recommendations from the Blue Ribbon Commission on America's Nuclear Future". Washington DC, 2012.
42 Nuclear Waste Policy Act Amendments Act of 1987, H.R. 3430, 100th Cong. (1987–1988)
43 Blue Ribbon Commission, "Recommendations from the Blue Ribbon Commission on America's Nuclear Future". Washington DC, 2012.

nual fee revenues (approximately $750 million per year) and the unspent $27 billion balance in the fund created for the disposal was part of the financial loss from stopping Yucca Mountain.

According to the US Government Accountability Office (GAO),[44] the US has over 86,000 metric tons of spent nuclear fuel from commercial nuclear power plants. DOE is responsible for disposing of this waste in a permanent geologic repository but has yet to build such a facility. As a result, the amount of spent nuclear fuel stored at nuclear power plants across the country continues to grow by about 2,000 metric tons a year. Meanwhile, the federal government has paid billions of dollars in damages to utilities for failing to dispose of this waste in addition to the siting and investigation of Yucca Mountain as the repository and may potentially have to pay billions of dollars more in coming decades according to the GAO.[45]

Today, spent nuclear fuel, which is the used fuel removed from core of the nuclear powerplant, is one of the most dangerous and, in many cases, long-lasting substances created by humans. Today, commercial spent fuel is temporarily stored at powerplant sites; about 74 percent of it is stored in pools of water, and 26 percent has been transferred to dry storage casks. According to the GAO,[46] the US has no permanent disposal site for the nearly 70,000 metric tons of spent fuel currently stored in 33 states that continues to grow.

The amount of spent fuel stored on-site at commercial nuclear powerplants will continue to accumulate – increasing by about 2,000 metric tons per year and likely more than doubling to about 140,000 metric tons – before it can be moved off-site, because storage or disposal facilities may take decades to develop. In examining interim storage or permanent disposal options, GAO found that new facilities may take from 15 to 40 years before they are ready to begin accepting spent fuel. Once an off-site facility is available, it will take several more decades to ship spent fuel to that facility. This situation will be challenging because by about 2040 most currently operating reactors are scheduled to retire operations, and options for managing spent fuel, if needed to meet transportation, storage,

---

44 United States Government Accountability Office, *Commercial Spent Nuclear Fuel: Congressional Action Needed to Break Impasse and Develop a Permanent Disposal Solution.* GAO-21–603, September 2021.

45 United States Government Accountability Office, "*Commercial Spent Nuclear Fuel.*"GAO-21–603, September 2021.

46 United States Government Accountability Office, *Spent Nuclear Fuel: Accumulating Quantities at Commercial Reactors Present Storage and Other Challenges.* GAO-12–797, August 2012.

or disposal requirements, may be limited. The Congressional Research Service[47] outlined many of the costs from litigation by utilities for the federal government of not opening Yucca Mountain. This cost is in the billions.

While other countries experiences are less fraught with political and human rights issues as the US in siting a nuclear waste facility, the issue of procedural justice was clearly undermined by the lack of attention to the local context in dealing with the high-level nuclear waste generated by these powerplants. For this energy source, small or large, it was hard to have equal distribution of benefits and harms with nuclear waste disposal because the benefits of the energy are displaced from its waste disposal. The harms and costs of using this energy in terms of disposing of its wastes are being transferred to future generations. Continued building of nuclear power plants without a waste disposal site is reckless public policy. In addition, it is hard to understand the recasting of nuclear energy as a solution to climate change when it creates other long-lasting concerns outside of a carbon neutral model. Thus, the sustainability of this expensive energy source as well as its long-term impacts without a disposal facility make is a questionable alternative to fossil fuels.

Understanding Uranium from the Dine Navajo Nation Culture Lens

As of current estimates, the Navajo Nation registered as the largest population of all the tribal nations pushing past the Cherokee Nation to nearly 400,000 people.[48] Among some 500 Indian tribes and 318 reservations recorded in the country by the 2000 US Census, the Navajo Nation is the home of the largest federally-recognized American Indian tribe (categories use by the US Census). The Diné people are believed to have originally migrated from western Canada and belonged to an American Indian group called the Athabascans. Some Athabascan bands first came into the southwest region of the US around the year 1300. Some settled in southern Arizona and New Mexico and became the different Apache tribes. Apache languages sound very much like Diné. By the year 1700, Diné people were living in northern

---

47 U.S. Congressional Research Service. *The Radiation Exposure Compensation Act (RECA): Compensation Related to Exposure to Radiation from Atomic Weapons Testing and Uranium Mining.* R43956, Jan. 13, 2021.
48 Simon Romero, "Navajo Nation Becomes Largest Tribe in the U.S After Pandemic Enrollment Surge". *New York Times Company*, May 21, 2021. Available online at: https://www.nytimes.com/2021/05/21/us/navajo-cherokee-population.html

Arizona, New Mexico, southern Colorado and Utah. They gave their land the name of *Dinétah* which resides within the four scared mountains in this Four Corner region. It is these sacred mountains that hold their creation and traditions. The Diné believe that they travelled through three worlds (Black, Blue and Yellow) to get to the current world which is called White.[49] The "Navajo Reservation" was established with the signing of the Treaty of 1868.[50] At that time, it was quite small, covering a territory containing Fort Defiance, Chinle, and Shiprock. Since then, the Diné lands expanded significantly to 27,000 acres.

The understanding of the sovereignty of the Navajo in legal terms is complex in US law. It dates to what has been called the Marshall Trilogy with three cases decided by the US Supreme Court under Chief Justice John Marshall, for whom it was named. These foundational doctrines have caused significant harm to the Diné people. The Navajo Nation in the US is not considered an independent state, as that term is used to define nation-states under international law. Instead, the US continues to describe their political status as a domestic dependent nation (semi-sovereign) with tribal sovereignty (keep in mind that the notion of a nation-state is a Western colonial construct).[51] The Diné paradigm of what the US refers to as the Navajo Nation is a world view based on Sąʼáh Naagháí Bikʼeh Hózhǫ́ǫ́ (SNBH). These are the principles of the Diné philosophy. SNBH is linked to all aspects to the Diné way of life. It represents a four-part planning and learning process. [52] The four tenets are thinking, planning, living, and assurance.[53] This is linked to sets of four that the Diné observe throughout life. For instance, this is akin to the four cardinal directions of east, west, south, north. East and north represent masculinity and west and south represent femininity. Another example is the sacred minerals of white shell, turquoise, abalone, and black jet or the four parts of the day being dawn, day, sunset, night, and the four seasons.[54] To the Diné,

---

49 Tom Souksavanh Koeovorabouth, "Reaching Back to Traditional Teachings: Diné Knowledge and Gender Politics". *Genealogy (Basel)* 5, no. 4 (2021): 95–104.

50 Peter Iverson and Roessel Monty, *Diné: a History of the Navajos*. 1st ed. Albuquerque: University of New Mexico Press, 2002.

51 Justice Raymond Austin, "Diné Sovereignty, a Legal and Tradition Analysis". In Lloyd L. Lee and Jennifer Denetdale (eds.), *Navajo Sovereignty: Understandings and Visions of the Diné People*. Tucson, Arizona: The University of Arizona Press, 2017, 19–42.

52 Lloyd Lance Lee and Gregory Cajete, *Diné Perspectives: Revitalizing and Reclaiming Navajo Thought*. University of Arizona Press, 2014.

53 Lloyd Lance Lee and Gregory Cajete, *Diné Perspectives*. Louise Lamphere and Marilyn Verney, "Navajo Religious Traditions" In Lindsay Jones (ed.), *Encyclopedia of Religion*. Macmillan, 2005.

54 Robert S. McPherson, *Dinéjí Naʼnitin: Navajo Traditional Teachings and History.* Colorado: University Press of Colorado, 2012. Tom Souksavanh Keovorabouth, "Reaching Back to Traditional Teachings".

SNBH is tied to the notion of interconnected force and energies in life. These energies tend to be viewed as binary opposites and can be tagged to masculinity and femininity. To the Diné, humans embody both femininity and masculinity which are not separate or monolithic as in western cultures. For instance, nature is described in a binary way of the Earth as the mother and the sky father. Also, balancing the positive and negative forces in life are parts of the SNBH. It is the equilibrium of these energies that once achieved in symmetry can bring a satisfying life. Too much of any one force can be helpful both to the individual and the community.[55] In this manner, the Diné are similar to the teachings of Taoism with binary opposites. SNBH is a matrix of the interconnectedness of life. Each Diné achieves SNBH through language, land, cultural, knowledge, protocol, trade, and living a sustainable way of life. In the Diné culture, knowledge is passed down via oral histories and narratives to provide people with a deep understanding of the Diné society and our relationship with the Dinétah. The Diné creation story indicates that there is no dichotomy between the natural and the supernatural in the Diné religion. The goal of the Diné life is to a live in the condition of Hózhǫ́ which is balance and harmony.

Repeated colonization by the US brought significant loss for the Diné people. One example was in 1864, when there was a forced relocation of the Diné from their homelands to Bosque Redondo Reservation at Fort Sumner in New Mexico (over 350 miles on foot) which served as a internment camp. This is one of the most important events for the Diné people called the "Long Walk."[56] This was colonialism aimed at eliminating the Diné by physically removing them from the Dinétah. This event is one of the major events of the community that remains a significant trauma for the community related to social injustices. The tribal nations have consistently associated their disproportionate rate of distress with historical experience of European colonization is referred to as historical trauma. The major source of the disparities in mental health appears to be the disproportionally high rates of exposure to potentially traumatic stressor that routinely led to increased prevalence of post-traumatic stress. Some of the health disparities from influence include a history of genocide, boarding school's acculturation, experiences, that have led to unresolved historical trauma and its associated poor health outcomes. [57] However, for socially oppressed communities that lack equal access to opportu-

55 Lee and Cajete, *Diné Perspectives*, 6–7.
56 Charles River Editors, *Native American Tribes: The History and Culture of the Navajo.* Create-Space Independent Publishing Platform, 2013.
57 Donald Warne and Denise Lajimodiere, "American Indian Health Disparities: Psychosocial Influences: American Indian Health Disparities". *Social and Personality Psychology Compass* 9, no. 10 (2015): 567–579.

nities and have a history of exploitation, survival and psychological, resilience can come in the form of survivance which requires a resistance to oppression.[58]

In 1968, the Navajos signed the Treaty of 1868 to return to their homeland and four scared mountains.[59] In 1934 the US government passed the Indian Reorganization Act to provide a provision for the making of the Native Nations. Under this legislation by the US government, the Navajo Nation government was formed which included the appointment of a tribal presidents, vice president, and 88 council delegates that represented 100 chapters of the tribe. This was a westernizing event for the Diné people. Further additional weakening of the Diné culture came later from years of sending children to boarding schools in the US for education since they were lacking at the Nation. This further entrenched the western Christian values into the Diné culture via assimilation that impacted their view of each other that continues today. The Western education system requires us to separate the religious from the secular. Diné people prefer to maintain their spirituality in a holistic perspective that leads to peace and harmony. Their relationships are based on the family like a clan system. The Diné people function as large extended family and it holds them together and is the basis for interactions in the culture.[60] This wisdom is often transferred in oral narrative from the elders to others. However, there are disconnects that exist today for the Diné in the transfer of this knowledge from the colonial interrupts to their culture. One study on the intergenerational understanding of the Diné found that while elders had the greatest in-depth understanding of the Hózhǫ́ (cultural wisdom) followed by the adolescents, the adults were the least understanding of it.[61] This could be a result from the years of colonialism and interruption to the Diné way of life.

The Diné philosophy of the mining of uranium is a disruption of the balance of the Earth and sky. One of the most powerful voices of the view of uranium mining comes from the Diné themselves using oral history research.[62] Many Diné refer to

---

58 Lucio Cloud Ramirez and Phillip L Hammack, "Surviving Colonization and the Quest for Healing: Narrative and Resilience Among California Indian Tribal Leaders". *Transcultural Psychiatry* 51, no. 1 (2014): 112–133.

59 Keovorabouth, "Reaching Back to Traditional Teachings", 5.

60 Herbert John Benally, "Spiritual Knowledge for a Secular Society: Traditional Navajo Spirituality Offers Lessons for the Nation". *Tribal college III*, no. 4 (1992): 19–25.

61 Michelle Kahn-John, Terry Badger, Marylyn Morris McEwen, Mary Koithan, Denise Saint Arnault, and Tara M Chico-Jarillo, "The Diné (Navajo) Hózhó Lifeway: A Focused Ethnography on Intergenerational Understanding of American Indian Cultural Wisdom". *Journal of Transcultural Nursing* 32, no. 3 (2021): 256–265.

62 Doug Brugge, Timothy Benally, and Esther Yazzie-Lewis, *The Navajo People and Uranium Mining*. Albuquerque: University of New Mexico Press, 2006.

the uranium as the *"leetso"* or powerful yellow monster, or yellow dirt.[63] Hundreds of Navajo people worked in the open-pit underground mines. No one told them of the dangers of radiation and the fight to obtain compensation benefits was difficult under RECA. There is an ancient Diné belief that humans should not dig deep into the Earth. Many of the Diné people supported the WW2 efforts as soldiers or code talkers. When the Manhattan Project, the secret nuclear weapons program of the defense department of WW2, needed uranium it went to the only domestic supply of uranium it was aware of which was the former vanadium mines on or near the Diné lands. Vanadium was used to strengthen steel and uranium was a by-product. Most of the tonnage of uranium was shipped from foreign sources with only 15% of the ore from the US. However, between 1943–1945, an estimated 44,000 pounds of uranium were secretly recovered from the Vanadium Corp of America for the Manhattan Project. Despite the Diné land source, most of the uranium came from the Belgian Congo and Canada.[64] Over 3,000 Navajo miners blasted raw uranium from the ground as ore, then processed it into yellowcake as part of the precursor materials needed. Although there was evidence in the 1950s of the health risks associated with uranium mining, the miners were never informed and not provided any occupational protection gear for the work.

Often the environmental justice researchers and climate activists do not acknowledge the intersection of tribal sovereignty and environmental justice in the context of colonialism. The environmental justice movement fails to address the issue of community self-determination, potentially leading to an uneasy relationship between a struggling tribal nation and environmental justice activists.[65] Environmental justice scholars tend to focus on the more recent issues of contamination, while the Diné experience links back to the colonization of their people with the arrival of the European man. The close relationship between environmental racism and settler colonialism which included the extraction of resources, labor, and goods in addition to taking land that was already home to indigenous people has been termed a form of *"wastelanding."*[66] Meaning, Euro-Americans view lands particularly in a desert as almost disposable, or not holding value except for the resources that can be extracted and wastes can be discard on. It is a form of en-

---

63 Esther Yazzie-Lewis and Jim Zion, "Leets, the Powerful Yellow Monster A Navajo Cultural Interpretation of Uranium Mining". In Doug Brugge, Timothy Bennally, and Esther Yazzie-Lewis (eds.), *The Navajo People and Uranium Mining*. Albuquerque: University of New Mexico Press, 2006. 1–10.
64 Traci Brynne Voyles, *Wastelanding: Legacies of Uranium Mining in Navajo Country.* Minneapolis: University of Minnesota Press, 2015. 2–3.
65 Noriko Ishiyama, "Environmental Justice and American Indian Tribal Sovereignty: Case Study of a Land-Use Conflict in Skull Valley, Utah". *Antipode* 35, no. 1 (2003): 119–139.
66 Voyles, *Wastelanding*.

vironmental privilege rendering lands as worthless which can be used for discarding human wastes to avoid impacting those in power. The underlying economic notion is one of a classic economics negative externality. The adverse impacts from waste are not born by those who benefit. Like the wastelanding view of discounting land, the Diné people who are associated with those lands are discounted as a form of "othering" which is part of environmental and social racism. The Diné people were sacrificed for economic growth during the expansion of the US and protection of the US in post-WW2. Many uranium miners died before the RECA of 1990 from radiation exposure. The miner in 1947 entered the mines to extract uranium ore and had no ventilation, drank the water from the hot puddles on the mine floor, and ate their food in the dust-filled mines. They brought their work clothes into the family homes and built their homes from the radioactive materials.[67]

Diné men were put in the position of accepting the mining jobs as a place to work near their homes which were some of the only jobs available. In 1949, miners were paid minimum wage or less of about .81 cents to $1 per hour.[68] Few Diné people spoke English and at the time of when the mining began. Uranium was not anything the Diné people had encountered before, as a result, there was no word for radiation in the Diné language.

Lessons Learned: Understanding the Future Consequences Nuclear Energy in the Race to Carbon Neutrality

The industrial prosperity post-WW2 in the US and the rise of nuclear technology has left The Diné and the US people with an unwanted and lethal legacy of radioactive waste.[69] Something has to be done to dispose of the large amounts of high-level nuclear waste already generated, and that continues today from nuclear energy plants still operating. The thought of creating new plants without a disposal solution in place is irresponsible. In fact, nuclear waste is proposed as an agent of environmental racism and oppression.[70] As the world races to carbon neutrality,

67 Rachel L. Spieldoch, "Uranium Is in My Body". *American Indian culture and Research Journal* 20, no. 2 (1996): 173–185.
68 Doug Brugge and Rob Goble, "The History of Uranium Mining and the Navajo People". *American Journal of Public Health* 92, no. 9 (2002): 1410–1419.
69 Ishiyama, "Environmental Justice", 123.
70 Andrea Boeckers, "Environmental Racism: Nuclear Waste as an Agent of Oppression?" *Across the Bridge: The Merrimack Undergraduate Research Journal* 1, Article 3 (2019).

the concepts of sustainability seem to be lost in the sense of urgency. For sure, climate change is a significant urgent matter. As renewable energy sources replace the fossil fuel plants, and battery storage technology develops, perhaps some lessons learned from the legacy to the Diné people in the extraction of uranium and the policy failure of disposal can help to inform the choices being made for the future dependency on the types of energy profile mix nation-states consider. It is concerning from a public policy standpoint when we consider intergenerational energy justice on what the impacts from nuclear waste disposal will be for the future. Today, the tribal nations oppose nuclear energy and what they view as toxic nuclear colonialism from its extraction and waste disposal.[71] Before we begin the race to nuclear energy dependency as a green solution to climate change, the other impacts to social justice, and negative externalities of the toxic burden need to be factored into this energy source decision. The energy justice movement is beginning to ask about intergenerational justice which is narrowly defined as climate issues by those advocating for nuclear energy. Europe and the other nuclear countries struggle with local communities not wanting to take the legacy of the nuclear energy waste.[72] Perhaps we should consider ways to solve the waste disposal and extraction issues before embarking on building new nuclear energy powerplants as a solution to climate change. Also, perhaps climate justice and energy justice need to be more inclusionary of the consequences to the Diné and others who bear the negative externality impacts without the benefits.

---

71 "Statement against Toxic Nuclear Colonialism by Tom Goldtooth, Executive Director, IEN: Indigenous Environmental Network." *Indigenous Environmental Network*, June 30, 2021. Available online at: https://www.ienearth.org/statement-against-toxic-nuclear-colonialism-by-tom-goldtooth-executive-director-ien/

72 Kalina Oroschakoff and Marion Solletty, "Burying the Atom: Europe Struggles to Dispose of Nuclear Waste". *Politico*, July 19, 2017. "Indigenous Anti-Nuclear Statement: Yucca Mountain and Private Fuel Storage at Skull Valley: Indigenous Environmental Network". *Indigenous Environmental Network*, November 1, 2021. Available online at: https://www.ienearth.org/indigenous-anti-nuclear-statement-yucca-mountain-and-private-fuel-storage-at-skull-valley/. "Environmental Justice Case Study: Accepting Money for Nuclear Waste in Skull Valley, Utah". *Skull Valley Justice Page*, accessed on January 27, 2022. Available online at: http://websites.umich.edu/~snre492/ibrown.html

Rosalia Gambino

Chapter II
Sustainable Waste? Waste Colonialism and the "Sustainable" Imaginary

Introduction

The idea of sustainable development advances the false claim[1] that protection of the environment ("sustainability") and perpetual economic growth ("development") can coexist. This is the primary narrative within the rapidly evolving physical and intellectual infrastructure that has kept us blissfully ignorant of the potential scales of violence and catastrophe that are ever-expanding. The United Nations claims that its 2030 Agenda for Sustainable Development, which fosters this narrative, "is a plan of action for people, planet and prosperity. It also seeks to strengthen universal peace in larger freedom" by way of "free[ing] the human race from the tyranny of poverty and want,"[2] but sustainable development does not stand in opposition to poverty and want nor can it benefit all of humanity. Perpetual raw resource extraction, production, and consumption, as key elements of economic growth and the sustainable development project, allow for the violence of waste colonialism.[3] A focus on colonization through waste demonstrates that colonialism is not only about the seizure of territory (though territory – space in which to dump waste in this case – is certainly a central feature); also significant is corporeal dispossession – the gradual seizure of control over a population's physical health, which can potentially lead to physical and cognitive impairment (which is certainly not "sustainable" for the human species). It is a slow and structural violence that places unequal burden on indigenous, racialized, and poor communities in the Global South.[4]

---

1 See Donnella H. Meadows, Dennis L. Meadows, Jørgen Randers, and William W. Behrens III, *The Limits to Growth.* New York: Universe Books 1972.
2 United Nations Department of Economic and Social Affairs, "Transforming our World: the 2030 Agenda for Sustainable Development", accessed on 4 May 2022. Available online at: https://sdgs.un.org/2030agenda
3 This is one form of pollution-as-colonialism. For a discussion on the subject, see Max Liboiron, *Pollution is Colonialism.* Durham and London: Duke University Press 2021.
4 In this chapter, "Global North" and "Global South" are understood as fluid categories; the distinction between the two typically establishes economic differences between states, but some researchers have argued that this distinction can also be made *within* countries. This often results in what

https://doi.org/10.1515/9783111081687-003

The first section of this chapter examines the available data on waste generation and its global flows, with particular emphasis on electronic and electrical waste (hereafter described as "e-waste"). The persistent flow of waste from the Global North to the Global South, despite the introduction of the Basel Convention on the Control of Transboundary Movements of Hazardous Wastes and their Disposal, is an expression of the violent and catastrophic practice of waste colonialism (including e-waste colonialism). The second section reviews recent data that demonstrate the harms caused by waste colonialism. It focuses on Guiyu, China and Lagos, Nigeria, two major recipients of much of the world's waste. The third section moves to examine "clean" energy technologies as contributors to e-waste colonialism and centers on solar photovoltaic panels (i.e. solar panels) since they are key in the sustainable development project. As the desire for sustainable development grows, but the desire to understand the global dynamics of this project does not, sustainable development will continue reenacting colonial violence. The fourth section explores a proposed solution – that of the "circular economy" – and asks whether it is indeed capable of resolving the e-waste problem. The conclusion calls for a reexamination of the sustainable development project.

The Violence and Catastrophe of Waste Colonialism

Waste continues to be generated at an increasingly accelerated rate because it is an inherent part of economic growth. Worldwide, 2.01 billion tons of municipal solid waste are generated annually, and by 2050, this number is expected to grow to 3.4 billion tons.[5] The United States alone produced 292.4 million tons of municipal solid

some refer to as the "global color line," which identifies the racial division between the Global North and Global South. This chapter understands the North/South distinction as both an indicator of wealth and racialized divisions and indeed recognizes that the Global South can exist within states typically identified as parts of the Global North. See Andrea Wolvers, Oliver Tappe, Tijo Salverda, and Tobias Schwarz, "Concepts of the Global South – Voices from Around the World", Global South Studies Center, University of Cologne (January 2015); Thomas E. Smith, *Emancipation Without Equality: Pan-African Activism and the Global Color Line*. Amherst and Boston: University of Massachusetts Press 2018. Marilyn Lake and Henry Reynolds, *Drawing the Global Colour Line: White Men's Countries and the Question of Racial Equality*. Melbourne: Melbourne University Press, 2008.
5 Silpa Kaza, Lisa Yao, Perinaz Bhada-Tata, and Frank Van Woerden, *What a Waste 2.0: A Global Snapshot of Solid Waste Management to 2050*. World Bank Group, 2018, 3.

waste in 2018, up from 208.3 million tons in 1990.[6] The World Bank Group states that "though they account for 16 percent of the world's population, high-income countries generate about 34 percent, or 683 million tonnes, of the world's waste."[7] The data on e-waste, which is part of the broader category of municipal solid waste, likewise shows an upward trend. The 53.6 million metric tons of e-waste (excluding solar panels) that was generated worldwide in 2019 is a 21 % increase from 2014.[8] This number is projected to reach 74.7 million metric tons by 2030.[9] In 2013, the National Institute of Environmental Health Sciences wrote that e-waste was the "fastest-growing stream of municipal solid waste," the management of which "is a significant environmental health concern."[10] The latter is especially true for those in the Global South who are most burdened by the global generation of waste.

The growing volume of waste has not been followed by an increase in recycling practices. From 2010 to 2019, e-waste grew by 38 %, but less than 20 % was recycled.[11] The other roughly 80 % wound up in a landfill or was informally recycled.[12] This informal recycling is often done by hand in the Global South, "exposing workers to hazardous and carcinogenic substances such as mercury, lead and cadmium...[which] contaminates soil and groundwater, putting food supply systems

6 United States Environmental Protection Agency, "National Overview: Facts and Figures on Materials, Wastes and Recycling", accessed on 14 July 2021. Available online at: https://www.epa.gov/facts-and-figures-about-materials-waste-and-recycling/national-overview-facts-and-figures-materials.
Municipal solid waste consists of everyday items such as electronics, food containers, product packaging, and clothing. This category does not include hazardous waste, agricultural and medical waste, or fossil fuel combustion waste, among others. See United States Environmental Protection Agency, "Wastes," Available online at: https://www.epa.gov/report-environment/wastes
7 Kaza et al., *What a Waste 2.0*, 3.
8 Vanessa Forti, Cornelis Peter Baldé, Ruediger Kuehr, and Garam Bel, *The Global E-waste Monitor 2020: Quantities, flows, and the circular economy potential.* Bonn/Geneva/Rotterdam: United Nations University/United Nations Institute for Training and Research – co-hosted SCYCLE Programme, International Telecommunication Union, and International Solid Waste Association, 2020, 23.
9 Forti et al, *Global E-waste Monitor 2020*, 13.
10 Paula T. Whitacre, "E-Waste Recycling in China: A Health Disaster in the Making?" *National Institute of Environmental Health Sciences*, July 2013. Available online at: https://www.niehs.nih.gov/research/programs/geh/geh_newsletter/2013/7/articles/ewaste_recycling_in_china_a_health_disaster_in_the_making.cfm
11 United Nations, *The Sustainable Development Goals Report 2020*, 17.
12 United Nations Environment Programme, "UN Report: Time to Seize Opportunity, Tackle Challenge of E-Waste," 24 January 2019. Available online at: https://www.unenvironment.org/news-and-stories/press-release/un-report-time-seize-opportunity-tackle-challenge-e-waste

and water sources at risk,"[13] which indicates that workers are not only risking their own health; they are also endangering nearby communities. E-waste is full of toxic substances that can lead to adverse health impacts, such as cardiovascular disease, DNA damage, and cancer.[14] The UN's 2020 Sustainable Development Report identifies the problematic nature of this development, specifically as it affects low-income countries:

> In high-income regions, an e-waste management infrastructure exists. However, collection rates are, on average, substantially below 50 per cent. E-waste materials are often categorized as reusable goods and can also be exported to middle- and low-income countries. But in many of those countries, infrastructure is not yet developed or is inadequate to manage locally generated and illegally imported e-waste. The waste is mostly handled by the informal sector through open burning or acid baths, both of which pollute the environment and result in the loss of valuable and scarce resources. Moreover, workers and their children, who live, work and play on those sites, often suffer severe health effects.[15]

The UN claims that its 2030 Agenda "provides a *shared* blueprint for peace and prosperity"[16] but if populations in the Global North do not curb their consumption, and therefore their waste production, more and more populations in the Global South will be forced to *consume waste* to the detriment of their biological capacities. This is an expression of slow, structural violence.

The term "violence" is often understood as "the deliberate exercise of physical force against a person, property, etc." or "physically violent behavior or treatment,"[17] but scholars often challenge these all-too-simplistic definitions. Johan Galtung, for example, writes about *structural* violence, in which "there may not be any person who directly harms another person in the structure. The violence is built into the structure and shows up as unequal power and consequently as unequal life chances."[18] This violence is often imperceptible and leads to the erasure of the experiences of those on whom this violence is enacted. Rob Nixon similarly refers to "slow violence," the "delayed destruction that is dispersed across time and

---

13 United Nations Environment Programme, "UN Report: Time to Seize Opportunity".

14 World Health Organization, *Children and digital dumpsites: E-waste exposure and child health.* World Health Organization, 2021, xv.

15 United Nations, *The Sustainable Development Goals Report 2020*, 48.

16 United Nations, "The 17 Goals – History", accessed on 4 May 2022. Available online at: https://sdgs.un.org/goals. [Emphasis added by author.]

17 Oxford English Dictionary, "violence, *n.*," 1a, accessed on 4 May 2022.

18 Johan Galtung, "Violence, Peace, and Peace Research". *Journal of Peace Research* 6, no. 3 (1969): 171.

space"[19] and focuses on what he refers to as "environmental catastrophes"[20] that affect human biology over time. A catastrophe is commonly defined as an *event*, a *sudden* change, and "a subversion of the order or system of things,"[21] but it is rather the result of the violence that builds onto itself over time until it eventually becomes apparent.[22]

The introduction of carcinogenic or otherwise harmful materials into the environment, thereafter available for consumption by living beings, follows the gradual dispossession of a group's geographical space by the waste of another; thus begins the slow manifestation of the structural violence of the Global North-South divide. The environmental disaster that manifests as that which we can identify as waste colonialism is a result of the very system that is said to be subverted in a catastrophe. The system itself is a violence- and catastrophe-inducing machine.

The Basel Convention, which entered into force in 1992 after the discovery that toxic waste was imported to the developing world from abroad,[23] is the most significant international agreement regarding global waste management, but it did not end the violence of waste colonialism. The next subsection discusses the Convention's attempts to address the global flows of waste to the Global South and its ultimate failure to do so.

The Basel Convention and Waste Colonialism

The Basel Convention aims to mitigate potential damage to both human health and the environment produced by the international transport of waste, particularly to the "developing" world, which has been considered ideal for disposal because environmental awareness is less prominent and regulation and enforcement mechanisms are lacking.[24] The Convention states that "the most effective way of protecting human health and the environment from the dangers posed by [hazardous and other] wastes is the reduction of their generation to a minimum in terms of quan-

---

19 Rob Nixon, *Slow Violence and the Environmentalism of the Poor.* Cambridge and London: Harvard University Press, 2011, 2.

20 Nixon, *Slow Violence*, 2.

21 Oxford English Dictionary, "catastrophe, *n.*," 2a, 3a, 3b, 4, accessed on 4 May 2022.

22 See also Antonio Y. Vázquez-Arroyo, "How Not to Learn From Catastrophe: Habermas, Critical Theory and the 'Catastrophization' of Political Life". *Political Theory* 41, no. 5 (2013): 738–765, especially 742.

23 United Nations Environment Programme, "History of Negotiations of the Basel Convention", accessed on 4 May 2022. Available online at: http://www.basel.int/TheConvention/Overview/History/Overview/tabid/3405/Default.aspx

24 Ibid.

tity and/or hazard potential,"[25] yet the Global North has continued to generate waste that is exported to developing states; the only caveat here is that the importing (i.e. developing) nation had to grant the exporting (i.e. developed) nation permission to do so.[26] This was ostensibly in respect of other states' sovereignty,[27] yet minimally so. Moreover, the impact of the Convention depends on global adherence to it, and the United States, one of the largest producers of waste per capita,[28] has yet to ratify it.[29]

Though states have the option to reject the import of waste, some have accepted it because of the potential value of the component parts in the products previously discarded; this waste could then be used to generate production, as was the case in China, which, until 2018, was the largest importer of waste in the world – it had accepted about 70 % of the world's waste since 1997.[30] This is a widely accepted statistic, but since less than 20 % of the world's e-waste is delivered to formal recyclers, the locations of the other roughly 80 % is unknown; "only 41 nations compile e-waste statistics," and even these are unreliable as their "partial data can't keep up with the expansion of electronic devices into so many consumer categories."[31] Many of these electronics that end up in landfills or are informally recycled lead to poor health outcomes for those who search for waste with value. Without first addressing the structural issues that lead workers to the informal waste economy, the Basel Convention is bound to be ineffective in alleviating the harms of waste colonialism.

This Convention was the first attempt to resolve the unfair disadvantage of states in the Global South, but another attempt came in the form of the Ban Amendment to the Basel Convention, which entered into force on December 5th,

---

25 United Nations Environment Programme, *Basel Convention on the Control of Transboundary Movements of Hazardous Wastes and their Disposal* (Revised in 2019), 4, accessed on 4 May 2022.
26 Ibid., 14.
27 Ibid., 4, 13.
28 Kaza et al., *What a Waste 2.0*, 20.
29 United Nation Environment Programme, "Parties to the Basel Convention on the Control of Transboundary Movements of Hazardous Wastes and their Disposal," accessed on 4 May 2022. Available online at: http://www.basel.int/Countries/StatusofRatifications/PartiesSignatories/tabid/4499/Default.aspx
30 Laura Parker, "China's Ban on Trash Imports Shifts Waste Crisis to Southeast Asia". *National Geographic*, 16 November 2018. Available online at: https://www.nationalgeographic.com/environment/2018/11/china-ban-plastic-trash-imports-shifts-waste-crisis-southeast-asia-malaysia/
31 Brook Larmer, "E-Waste Offers an Economic Opportunity as Well as Toxicity". *The New York Times Magazine*, 5 July 2018. Available online at: https://www.nytimes.com/2018/07/05/magazine/e-waste-offers-an-economic-opportunity-as-well-as-toxicity.html

2019.[32] This Amendment "prohibits the export of hazardous waste from developed countries (OECD, EU member states, Liechtenstein) to developing countries."[33] Though quite a significant development, this Amendment does not include plastic, scrap metal, or paper waste, which could still be hazardous to health; plastic, for example, is constructed of toxic materials, such as bisphenol A (BPA) and brominated flame retardants, which can cause cardiovascular problems, neurological disorders, and the disruption of hormones that regulate the nervous and reproductive systems.[34] The Amendment also pertains only to those who agree to its terms; out of the 189 parties to the Basel Convention,[35] only 100 are party to the Ban Amendment.[36]

Even though the Basel Convention has been attempting to address e-waste issues since 2002,[37] it has not resolved (and likely will not resolve) the problem of waste colonialism. For example, between 2008 and 2012, the Basel Convention led to an e-waste Africa programme, "a comprehensive programme aiming to enhance the environmental governance of e-wastes and to create favorable social and economic conditions for partnerships and small businesses in the recycling sector in Africa."[38] Nonetheless:

> The management of hazardous wastes in most African countries is still characterised by lack of public awareness, poor political will, absence of legislation and standards, inadequate training and skills of relevant regulatory and enforcement officials, lack of funds, infrastructure, investments and technical equipment for recycling, treatment and sound disposal, lack

---

32 United Nations Environment Programme, "Entry into force of amendment to UN treaty boosts efforts to prevent waste dumping", 13 September 2019. Available online at: http://www.basel.int/Default.aspx?tabid=8120
33 Ibid.
34 Okunola A. Alabi, Kehinde I. Ologbonjaye, Oluwaseun Awosolu, and Olufiropo E. Alalade, "Public and Environmental Health Effects of Plastic Wastes Disposal: A Review". *Journal of Toxicology and Risk Assessment* 5, no. 1 (2019): 8–9.
35 United Nations Environment Programme, "Parties to the Basel Convention".
36 United Nations Environment Programme, "Amendment to the Basel Convention on the Control of Transboundary Movements of Hazardous Wastes and their Disposal", accessed on 4 May 2022. Available online at: http://www.basel.int/Countries/StatusofRatifications/BanAmendment/tabid/1344/Default.aspx
37 United Nations Environment Programme, "Overview", accessed 4 May 2022. Available online at: http://www.basel.int/Implementation/Ewaste/Overview/tabid/4063/Default.aspx
38 United Nations Environment Programme, "Overview", accessed 4 May 2022. Available online at: http://www.basel.int/Implementation/Ewaste/EwasteAfricaProject/Overview/tabid/2546/Default.aspx

of adequate data on waste flows, and inadequate international cooperation towards ESM [environmentally sound management].[39]

The Basel Convention Coordinating Centre for Training and Technology Transfer for the African Region published a 2020–2023 business plan that aims toward the resolution of these issues,[40] but the damage requires coordination between many actors and levels of governance.

The World Health Organization (WHO) released its first comprehensive report on the e-waste problem in 2021, in which it acknowledges that, in addition to the "growing e-waste dumps of Africa, Asia and Latin America...some developed economies" also serve as repositories of e-waste that expose those who informally work in these "dumps" to dangerous chemicals.[41] The report argues "that more assertive action is needed by the global, national and local health sectors in order to place the e-waste issue at the centre of health agendas and stimulate more effective and binding actions by e-waste importers, exporters and governments."[42] The results of these actions are forthcoming. For now, much of the world's waste resides in major waste sites in the Global South. The next section examines two such sites.

Waste Colonialism in Practice

China: Guiyu

One hot spot for the world's e-waste in particular is Guiyu, a town in Guangdong Province in China. In 2011, there were about 150,000 workers processing 1.5 million tons of e-waste each year.[43] In 2013, it was found that more than 80 % of families in Guiyu were involved in e-waste recycling, with over 5,500 e-waste businesses employing over 30,000 people.[44] Much of this e-waste was processed by hand and

---

39 Percy C. Onianwa, *Business Plan / Work Plan For the Period (01 January 2020–31 December 2023)*, Basel Convention Coordinating Centre for Training and Technology Transfer for the African Region, in Nigeria (BCCC-Africa), 5.
40 Onianwa, *Business Plan*.
41 World Health Organization, *Children and digital dumpsites*, vi.
42 World Health Organization, *Children and digital dumpsites*, xvii.
43 Chien-ming Chung, "China's E-Waste City", *The Virginia Quarterly Review* (Spring 2011), 86.
44 Janet Kit Yan Chan and Ming H. Wong, "A review of environmental fate, body burdens, and human health risk assessment of PCDD/Fs at two typical electronic waste recycling sites in China". *Science of the Total Environment* 463–464 (2013): 1113.

has placed both the population and its environment at risk of pollution.[45] Water and soil in the area are unsafe because of plastic residue, chromium, tin, and other heavy metals, and nearly all workers who process e-waste have respiratory illnesses.[46] Exposure to toxic substances also affects workers' thyroid function and causes DNA damage.[47] Workers are exposed to high levels of polybrominated diphenyl ethers (PBDEs), which can additionally cause cancer and reproductive and neurological harm, from open burning sites in Guiyu; PBDE levels "were some of the highest found in any environmental medium...and more than 16000 times higher than those found in soil samples in a distant reservoir used as control site."[48] PBDEs, along with a host of other elements toxic to the human body[49] that are prevalent in Guiyu, have severe implications for those working within the informal e-waste industry, as well as their families, at all levels of human development.

Children, for example, are in great danger of absorbing harmful chemicals that place them at risk for chronic health conditions. Children who live in Guiyu "have some of the highest lead levels in the world"[50] and are exposed to other toxic elements that affect their development, such as mercury, aluminum, and persistent organic pollutants (POPs).[51] Exposure to e-waste toxins also lead to adverse outcomes for pregnant woman, including spontaneous abortions and stillbirths.[52] Fetuses and children "are particularly vulnerable to several known and suspected

---

45 See Chung, "China's E-Waste City", 84–95. Xin Tong and Jici Wang, "The shadow of the global network: e-waste flows to China". In Catherine Alexander and Joshua Reno (eds.), *Economies of Recycling: The Global Transformation of Materials, Values and Social Relations.* Zed Books, 2012, 98–116. I.M.S.K. Ilankoon, Yousef Ghorbani, Meng Nan Chong, Gamini Herath, and Thandazile Moyo, "E-waste in the international context – A review of trade flows, regulations, hazards, waste management strategies and technologies for value recovery", *Waste Management* 82 (2018): 258–275.
46 Chung, "China's E-Waste City", 86.
47 Chiara Frazzoli, Orish Ebere Orisakwe, Roberto Dragone, and Alberto Mantovani, "Diagnostic health risk assessment of electronic waste on the general population in developing countries' scenarios", *Environmental Impact Assessment Review* 30 (2010): 390.
48 Frazzoli et al., "Diagnostic health risk assessment", 391–392.
49 Frazzoli et al., "Diagnostic health risk assessment", 391–392.
50 Chung, "China's E-Waste City", 86. See also World Health Organization, *Children and digital dumpsites,* 49.
51 Frazzoli et al., "Diagnostic health risk assessment", 389–391.
52 Kristen Grant, Fiona C. Goldizen, Peter D. Sly, Marie-Noel Brune, Maria Neira, Martin van den Berg, and Rosana E. Norman, "Health Consequences of Exposure to E-waste: A Systematic Review". *Lancet Global Health* 1 (2013): 353.

developmental neurotoxicants in e-waste."[53] The physical impact of the absorption of toxic materials harms the individuals that work in waste sites, but there is also a generational impact, both on the cycle of poverty that has become increasingly difficult to escape and on the biological development of future generations. Currently, "preliminary significant data are available for proving the transgenerational exposure" to toxic materials in Guiyu,[54] which suggests that the slow violence of waste colonialism may not be fully grasped for years to come.

This concentration of toxic elements in the environment has resulted from an unregulated waste recycling system,[55] which has changed a bit since 2013, with much of the e-waste moved to an industrial park, but workers are still at risk of physical harm. Those who work inside this park are still disassembling waste with simple tools and basic gear in poorly ventilated rooms.[56] Workers therefore continue to consume toxins even as the waste has become more concentrated and hidden away. The changes have also maintained the poverty of the workers, who have "become poorer, while their labor and rent payments enrich the government and line the pockets of the local businessmen, who can be spotted cruising around the park in luxury sedans."[57] Since the opening of this industrial park, tens of thousands of e-waste workers have been laid off and many have been forced to leave because of declining employment opportunities.[58] There are currently no measures in place to improve this situation.

---

53 Aimin Chen, Kim N. Dietrich, Xia Huo, and Shuk-mei Ho, "Developmental Neurotoxicants in E-Waste: An Emerging Health Concern," *Environmental Health Perspectives* 119, no. 4 (April 2011), 431. See also World Health Organization, *Children and digital dumpsites*, 23–28.

54 Frazzoli et al., "Diagnostic health risk assessment", 395. See also Martin Lappé, Robbin Jeffries Hein, and Hannah Landecker, "Environmental Politics of Reproduction", *Annual Review of Anthropology* 48 (2019): 135.

55 Frazzoli et al., "Diagnostic health risk assessment", 392.

56 Davor Mujezinovic, "Electronic Waste in Guiyu: A City under Change?" Environment & Society Portal, *Arcadia* (Summer 2019), no. 29. Rachel Carson Center for Environment and Society. Available online at: https://www.environmentandsociety.org/arcadia/electronic-waste-guiyu-city-under-change.

57 Mujezinovic, "Electronic Waste in Guiyu".

58 Kun Wang, Junxi Qian, and Shenjing He, "Global destruction networks and hybrid e-waste economies: Practices and embeddedness in Guiyu, China". *EPA: Economy and Space* 54, no 3 (2022): 548.

Nigeria: Lagos

These harmful dynamics also exist in the lives of informal waste workers in Nigeria, another major e-waste hub.[59] Nigeria's e-waste is generated both at home and abroad, from the European Union, the UK, the US, and elsewhere.[60] In both 2015 and 2016, approximately 60,000–71,000 tons of e-waste were imported into Nigeria through two main ports in Lagos.[61] Nigeria is thus widely accepted as a legitimate site of disposal. Much of this e-waste ends up at the Alaba International Market in Ojo, which is "the largest market for used and new electronics and electrical equipment in West Africa" and contains an informal e-waste dismantling and recycling site.[62] Here, "manual dismantling of electronics to recover metals such as copper, aluminium and other precious metals as well as open burning of some electronic components and wire cables are carried out."[63] The soil at this informal e-waste dismantling site has been found to have an "extremely high level of pollution" when compared to a Lagos State University campus used as a control site, with cadmium 12 times, chromium 123 times, copper 215 times, and lead 102 times higher than the control.[64] Sites such as this place workers at risk of heavy metal accumulation in their bodies and the development of respiratory problems, cancers, and other harmful diseases, but many nonetheless rely on this work for subsistence.

There are as many as 100,000 people working in the informal e-waste recycling sector who then collect and dismantle electronics by hand and dump or burn material that does not have economic value.[65] This exposes workers to hazardous materials, such as cadmium and mercury, and they "commonly suffer respiratory and dermatological problems, eye infections and lower than average life expectancy."[66] The e-waste that enters the country from abroad includes lead, hy-

---

59 "Lagos" is often used to refer to both Lagos, the city, and Lagos State. In this section, I refer to a few cities within Lagos State.
60 Irene Galan, "Dark skies, bright future: overcoming Nigeria's e-waste epidemic", *United Nations Environment Programme*, 7 August 2019. Available online at: https://www.unep.org/news-and-stories/story/dark-skies-bright-future-overcoming-nigerias-e-waste-epidemic. Irene Galan, "Nigeria turns the tide on electronic waste", *United Nations Environment Programme*, 19 June 2019. Available online at: https://www.unep.org/news-and-stories/press-release/nigeria-turns-tide-electronic-waste
61 Forti et al., *The Global E-waste Monitor*, 2020, 71.
62 Khadijah A. Isimekhai, Hemda Garelick, John Watt, and Diane Purchase, "Heavy metals distribution and risk assessment in soil from an informal E-waste recycling site in Lagos State, Nigeria". *Environmental Science and Pollution Research* 24 (2017): 17207.
63 Isimekhai et al., "Heavy metals distribution", 17207.
64 Isimekhai et al., "Heavy metals distribution", 17207, 17211.
65 Galan, "Nigeria turns the tide on electronic waste".
66 Galan, "Nigeria turns the tide on electronic waste".

drochlorofluorocarbons from refrigerators and air conditioners, and plastic components that are all harmful to human biology.[67] Workers are also at risk of sustaining injuries, including cuts which increase the risk of acquiring infectious diseases.[68] This is often the price that these communities pay to earn their livelihoods, often beginning at a young age.

Even though these workers are often exposed to toxic materials, they often lack personal protective equipment.[69] Those who do have access to boots and gloves do not change these items regularly and thus become more susceptible to harm.[70] Most of these workers, 86.7% of which are between 19 and 30 years old,[71] nonetheless work between 10 to 14 hours a day.[72] This work therefore prevents these individuals from seeking alternative opportunities for employment, but most earn more than the country's minimum monthly wage of ₦18,000 ($58.7).[73]

Similar experiences occur throughout Nigeria, including in Olusosun, the largest dumpsite in Lagos State, which covers 100 acres,[74] and Bauchi city in the northeast.[75] Searching and transforming waste into cash is a legitimate source of income for many, but the risk of negative health outcomes due to this labor needs more attention. Poor regulatory frameworks and enforcement in Nigeria[76] allow these workers to continue risking their lives for their livelihoods. There is hope that the circular economy, discussed in the final section of this chapter, will positively impact the lives of these workers,[77] but it does not reduce the physical impact of direct and persistent exposure.

---

67 Galan, "Dark skies, bright future".

68 Ahmed Fate Ali and Farouk Idi Yusuf, "Prevalence of Injuries Among Waste Pickers. A Case Study in Nigeria". *Detritus* 17 (2021): 90.

69 Isaac Jacob Omosimua, Olurinola Isaiah Oluranti, Gershon Obindah, and Aderounmu Busayo, "Working Conditions and Career Aspirations of Waste Pickers in Lagos State". *Recycling* 6, no. 1 (2020): 2.

70 Omosimua et al. , "Working Conditions and Career Aspirations of Waste Pickers in Lagos State", 9.

71 Toochukwu Chibueze Ogwueleka and Naveen B. P., "Activities of informal recycling sector in North-Central, Nigeria". *Energy Nexus* 1 (2021): 5.

72 Omosimua et al., "Working Conditions and Career Aspirations of Waste Pickers in Lagos State", 10.

73 Omosimua et al., "Working Conditions and Career Aspirations of Waste Pickers in Lagos State", 10.

74 Galan, "Dark skies, bright future".

75 Ali and Yusuf, "Prevalence of Injuries Among Waste Pickers".

76 Ilankoon et al., "E-waste in the international context", 264.

77 Galan, "Nigeria turns the tide on electronic waste".

These global waste flows that culminate in chosen repositories of waste are acts of corporeal dispossession because they establish a dependency on individuals who feel that they must come into daily contact with harmful materials to survive, which ironically destroys the ability to make decisions that are necessary for their biological survival. Galtung argues that structural violence "shows up as unequal power and consequently as unequal life chances,"[78] and the life chances of those who feel that they must work amongst others' waste are certainly diminished, even if the harms manifest slowly. The persistent choice to dump waste in the Global South is itself a symptom of (neo)colonial relationships in which the Global North uses its economic power to establish these dependencies and maintain established hierarchies. Even if the global flows of waste are no longer permitted, the extensive waste that has piled up in these developing states, and the health problems that have accumulated, will remain symbolic of these relations.

How Might "Clean" Energy Technology Contribute to Waste Colonialism? The Case of Solar Panels

Technologies that are both formed from natural materials and used to harness nature's energy – from the sun, wind, or water, for example – are understood as benevolently "clean"[79] as opposed to their "dirty" fossil fuel counterparts, even though fossil fuels are also of the earth. Fossil fuels result in further warming of the planet and anthropogenic climate change, while "clean" sources are thought to mitigate climate change and ensure the future survival of humankind. The designation of "clean" or "dirty" is therefore dependent on the method with which energy is harnessed and ways in which materials are transformed. But when "clean" technologies (e. g. solar panels; electric vehicles; wind turbines) are destroyed or reach the ends of their lifespans, they too become "dirty" as they transform into e-waste. They must then be recycled or disposed of, at which point they, too, have the potential to perpetuate waste colonialism and contribute to physical impairment in indigenous, racialized, and poor communities around the globe.

---

78 Johan Galtung, "Violence, Peace, and Peace Research", *Journal of Peace Research* 6, no. 3 (1969): 171.

79 The terms "clean energy" and "renewable energy" are often used interchangeably, and although they are used to denote energy sources that can presumably halt anthropogenic climate change, they do not share a definition. "Clean energy" sources do not emit greenhouse gases (which means that nuclear energy is technically "clean"), while "renewable energy" sources can replenish themselves. "Clean" energy is not always renewable.

We can see the start of this by examining the case of solar photovoltaic (PV) panels (known colloquially as simply "solar panels"), which have become major sources of "clean" and "renewable" energy. In the off-grid solar market alone, almost 3.5 million units of solar products were sold between January and June 2021.[80] Solar panels are expected to last about 30 years,[81] but increasing reliance on solar energy means that these numbers will multiply in the coming decades, and the potential hazards will multiply as well. Roughly 37 % of waste is disposed of in a landfill and 33 % is openly dumped.[82] The International Renewable Energy Agency and the International Energy Agency projected in 2016 that 60–78 million metric tons of solar PV panel waste will accumulate by 2050,[83] which has great potential to negatively impact the health of populations living near landfills and dumping sites in which the waste will come to rest. Improper disposal is problematic because of the hazardous materials used in their production, such as cadmium, which is highly toxic.[84]

With the increasing reliance on solar PV panels, the potential waste produced by lithium-ion batteries, which are necessary to store energy harnessed by the panels,[85] also needs to be considered. To transition to 100 % "clean" energy to ensure that the global temperature rise remains below 1.5 degrees preindustrial levels to meet the goals of the Paris Agreement,[86] 280 % of the earth's lithium reserves (which refers to the portions that are commercially viable to extract) would need to be mined.[87] This would likely continue a form of waste colonialism that takes place at the beginning of the supply chain, and, without proper recycling mechanisms in place, has the potential to enact waste colonialism at the end of

80 Global Off-Grid Lighting Association, *Global Off-Grid Solar Market Report: Semi-Annual Sales and Impact Data* (January-June 2021), 10.
81 International Renewable Energy Agency and the International Energy Agency Photovoltaic Power Systems, *End-of-life Management: Solar Photovoltaic Panels* (June 2016), 11.
82 Kaza et al., *What a Waste 2.0*, 5.
83 International Renewable Energy Agency and the International Energy Agency Photovoltaic Power Systems, *End-of-life Management*, 20.
84 Amalesh Dhar, M. Anne Naeth, P. Dev Jennings, and Mohamed Gamal El-Din, "Perspectives on environmental Impacts and a land reclamation strategy for solar and wind energy systems". *Science of the Total Environment* 718 (2020) 134602: 6.
85 "Lithium-ion batteries need to be greener and ethical", *Nature* 595 (1 July 2021): 7.
86 Article 2 of the Paris Agreement states that the global response to climate change includes "holding the increase in the global average temperature to well below 2 °C above pre-industrial levels and pursuing efforts to limit the temperature increase to 1.5 °C above pre-industrial levels, recognizing that this would significantly reduce the risks and impacts of climate change." United Nations, *Paris Agreement* (2015).
87 Elsa Dominish, Sven Teske, and Nick Florin, *Responsible minerals sourcing for renewable energy* (Institute for Sustainable Futures, University of Technology Sydney 2019), i and 21.

the supply chain. According to a 2011 United Nations Environment Programme (UNEP) report, less than 1% of lithium is recycled,[88] which means that most of it likely ends up as unprocessed e-waste.

Lithium also contributes to exploitative relationships in what is often referred to as the "Lithium Triangle," which includes Chile, Argentina, and Bolivia and contains over 75 percent of the world's supply of lithium.[89] In these countries, lithium mining pollutes water supplies, air, and soil that humans and animals rely on.[90] Pollution thereby affects the indigenous communities in these states, including the Aymara, Quechua, and Atacameño communities in Chile.[91] With the rise of solar PV panels as trusted sources of "clean" energy, the slow and structural violence of waste colonialism will likely persist as the Global North continues to benefit at the expense of the Global South. Perhaps if investments are made in safe recycling practices that limit the amount of lithium and other elements that need to be extracted, the situation might improve; as discussed in the next section, much hope lies in the future of recycling, in what is known as the "circular economy."

Can the Circular Economy Resolve the Global E-Waste Problem?

In January 2019, the Platform for Accelerating the Circular Economy and the E-waste Coalition, consisting of seven UN entities including the International Labor Organization and UNEP, released a report that addresses the rapid generation of e-waste as a global challenge, but the report argues that this also presents an opportunity for job development and economic growth. This document, *A New Circular Vision for Electronics*, forecasted that "by 2021 the annual total volume [of e-waste was] expected to surpass 52 million tonnes," and by 2040, "carbon emissions from the production and use of electronics...will reach 14% of total emissions."[92] Instead of focusing on the destructive capacity of increasing production,

---

88 International Resource Panel, *Recycling Rates of Metals: A Status Report* (United Nations Environmental Programme 2011), 18–19.

89 Samar Ahmad, "The Lithium Triangle: Where Chile, Argentina, and Bolivia Meet". *Harvard International Review* (Winter 2020): 51.

90 Ibid., 53.

91 Elena Giglio, "Extractivism and its socio-environmental impact in South America. Overview of the 'lithium triangle'", *América Crítica* 5, no. 1 (2021): 51.

92 Platform for Accelerating the Circular Economy and World Economic Forum, *A New Circular Vision for Electronics: Time for a Global Reboot* (January 2019), 10.

this report envisions what it refers to as a "circular economy," a "system in which all materials and components are kept at their highest value at all times, and waste is designed out of the system."[93] They believe that this type of economy is possible if new business and recycling practices are embraced.[94]

The WHO likewise believes that this is a necessary path that must be taken to resolve the e-waste problem but understands that the implementation of a circular economy would require a number of steps be taken first. "Healthier communities and reduction of costs can be obtained," states the WHO, "from more sustainable and health-focused e-waste policies that stimulate transition to a circular waste economy."[95] This requires a great deal of effort at local, national, and international levels, and capacity for this transition is currently lacking, even in "developed" states. In 2019, only 17.4% of the 53.6 million metric tons generated worldwide was "officially documented as properly collected and recycled,"[96] which indicates that drastic changes need to be made to handle the 74.7 million metric tons projected to be generated globally by 2030.[97] While the "e-waste sector could create millions of jobs worldwide,"[98] these jobs must be generated, managed for efficiency, and scrutinized for their safety for all biological life.

Other international bodies, including the United Nations Institute for Training and Research, the International Telecommunication Union, and the International Solid Waste Association, have also written in support of this circular economy. These organizations have collaborated on *The Global E-waste Monitor 2020* report in which they state the potential of e-waste in not only supporting the economy, but mitigating negative impacts on the environment. They state that "due to issues relating to primary mining, market price fluctuations, material scarcity, availability, and access to resources, it has become necessary to improve the mining of secondary sources and reduce the pressure on virgin materials. By recycling e-waste, countries could at least mitigate their material demand in a secure and sustainable way."[99] With the emphasis on the economic need for "resources" – which is an accepted term for identifying the environment as existing with the sole purpose of

---

93 Platform for Accelerating the Circular Economy and World Economic Forum, *A New Circular Vision for Electronics*, 16.
94 Platform for Accelerating the Circular Economy and World Economic Forum, *A New Circular Vision for Electronics*, 16.
95 World Health Organization, *Children and digital dumpsites*, xviii.
96 Forti et al., *The Global E-waste Monitor* 2020, 9.
97 Forti et al., *The Global E-waste Monitor* 2020, 13.
98 Platform for Accelerating the Circular Economy and World Economic Forum, *A New Circular Vision for Electronics*, 18.
99 Forti et al., *The Global E-waste Monitor* 2020, 58.

supporting humankind – is it hard to imagine that "reduc[ing] the pressure on virgin materials" is about respecting the environment at all.

Nonetheless, the UN and its member states adopted the 2030 Agenda for Sustainable Development in 2015, which includes goals such as ending poverty, ensuring clean water and sanitation, implementing "clean" energy, and building sustainable communities. A circular economy may work toward these goals, but it does not address the e-waste that has already accumulated around the world. How can we recycle e-waste in poor cities that do not have the tools for large-scale recycling? And what will happen to those who depend on this work for survival if there are formal processes put in place to recycle these materials? A circular economy is an ambitious project, but there are many structural issues that it leaves unaddressed. If informal workers are not widely integrated into this circular economy, projects that contribute to the Sustainable Development Goals will simply reinforce the Global North/South divide that is perpetuated by waste colonialism.

The United Nations' Sustainable Development Goals (SDGs) are meant to improve the well-being of states all over the world, but they too promote growth in a Western sense, which perpetuates the cycle of exploitation and dependency rather than alleviates it. These goals – these ideals – are not meant to conflict with global capitalism as we know it. The UN supports Industry, Innovation, and Infrastructure that is resilient and sustainable (goal 9) in a way that promotes energy efficiency,[100] a move towards Responsible Consumption and Production (goal 12), and the careful management of Life Below Water (goal 13) by avoiding pollution in our waterways.[101] Each of these (though all SDGs are implicated here) is necessarily related to the production of waste and its destruction of the environment but does not call for a change in our economic models or culture of consumption.

Conclusion

This chapter has demonstrated that waste colonialism, and e-waste colonialism in particular, upholds the structures that enact slow and structural violence in the Global South. Proposed solutions to impending environmental destruction, including sustainable development and a circular economy, are currently far removed

---

100 United Nations, "Goal 9: Build resilient infrastructure, promote sustainable industrialization and foster innovation," accessed 4 May 2022. Available online at: https://www.un.org/sustaina bledevelopment/infrastructure-industrialization/.
101 United Nations, "Take Action for the Sustainable Development Goals," accessed 4 May 2022. Available online at: https://www.un.org/sustainabledevelopment/sustainable-development-goals/.

from the local contexts that are most affected. Even though the urgency that must necessarily be felt and practiced around this issue seems to be calling the attention of international organizations, market mechanisms and economic growth seem to be prioritized over enacting proper governance and enforcement mechanisms. Moreover, any action that takes place in the hope of mitigating damage tends to be prioritized as a matter of environmental protection rather than as a matter of health.

Harmful toxins from waste seep from waste sites to the surrounding environment in which sentient beings reside and are ultimately absorbed into the body. Impairments of the body may not be observable for many years after the introduction of toxins, which is a major reason that attributing responsibility to any actor has been so difficult. Much of the world's waste has so far been shipped from the Global North to the Global South, but with changes in international arrangements and the exponential rise in waste, nations in the North will likely begin to feel the weight of their waste sooner than they imagined they would.

Sustainable development will have to be redefined to consider both the economic system and the health of biological systems. The current definition is about maintaining practices in production and consumption rather than addressing community health, but by generating products for consumption, we are also generating waste for consumption. The idea of "waste" is constructed of narratives that are necessary in a capitalist world – they allow for the perpetuation of this dominant model – but they also distance us from the damage that it brings. The extent of production and consumption in the Global North has become part of the culture, which has led to the feeling of entitlement towards products and ensures that the process continues. To resolve the environmental issues at hand, we have so far attempted to find purportedly better and "greener" methods of consumption and production, but we still believe that we can do so at the rate that we have always been operating. As long as "development" and "sustainable development" continue to be the narratives that drive our ideas of "progress," waste will continue to damage health for the perceived advantage of economic growth. This is a reflection of the environmental ontology of domination that must be deconstructed as we realize that there is not enough space to dominate, or that, in reality, the waste that we produce is beginning to destroy us.

Svetlana Bokeriya
Chapter III
Security Challenges of the Climate Change in the Sahel Region

The impact of global warming on the natural ecosystem of the planet often leads to geopolitical changes that jeopardize the stability of politically fragile regions, such as the Sahel. Pressure on scarce natural resources reduces the potential for international organizations to manage the region and decrease the risk of conflict. Within the Sahel, climate change can be predicted with a high degree of accuracy, compared to other risk factors. Currently, the entire regional population feels the effects of climate change. Rising temperatures have harmed health, livelihoods, food security, water availability, and physical security. Although there is significant international awareness about climate change issues, existing mechanisms to address them remain largely at the national level, due to the lack of multilateral regulatory mechanisms and environmental laws. The absence of such a mechanism not only reduces the opportunities for effective long-term action, but also pushes regional states to unilaterally manage environmental issues and security impacts.

This article analyzes the climate change problem and security challenges in the Sahel region. The purpose of this research is to identify the key approaches of international organizations in addressing the intersection of climate and security challenges in the Sahel. The analysis shows the evolution of UN and AU approaches to these issues, reflects the security and environmental aspects they need to refine to reduce regional conflicts, with the aim of having a positive impact on resolving the region's migration crisis, terrorist threat, and reducing the overall number of regional conflicts.

Since the 1992 Brazil Earth Summit, climate change and its associated risks to life on Earth have pushed governments around the world to find ways to fight the threat. Policies were developed, state institutions were established, and programs have been launched to this end.[1] The analysis of international efforts shows that, in relation to the enormity and gravity of climate change, progress made by governments has been few and far between, limited and generally disappointing. Historical issuers of such responsibilities have failed to meet their financing commit-

---

1 Handl, G. "Declaration of the United Nations conference on the human environment (Stockholm Declaration), 1972 and the Rio Declaration on Environment and Development, 1992". *United Nations Audiovisual Library of International Law, 2012*, last modified April 25, 2022. Available online at: https://legal.un.org/avl/ha/dunche/dunche.html.Handl

https://doi.org/10.1515/9783111081687-004

ments and promises within the framework of common but differentiated responsibilities[2] African developing countries have engaged the biggest polluters in international negotiations, but progress has been painstakingly slow. Current international climate governance has resulted in scattered actions that have a direct binding effect on countries' capability to cope with climate change.

Obviously, the growing threat of climate change poses significant risks to Africa's ambitions to renew itself in a deeply dynamic way. The challenge of change is already too expensive for African states, and for this reason, nothing less than the dream of an African Renaissance is at stake.[3] Since the beginning of the 21st century, African states have improved their cohesion, speaking with one voice in international conventions and conferences, forging instrumental relations with developing countries and creating strategic economic partnerships with emerging states such as Brazil, China, and Russia. African states have also identified with multipolarity, as evidenced by Russia and South Africa's membership in BRICS. They have also formulated a common position under the Committee of African Heads of State and Government on Climate Change (CAHOSCC) at several recent conferences and preparatory meetings on climate change international negotiations. Likewise, African solidarity on environmental issues was demonstrated during the RIO + 20 conferences held in June 2012 in Rio de Janeiro.[4] Due to this collective unity, questions related to economic advancement and the establishment of an institutional framework for sustainable development have come to bear on Africa's solidarity and participation in such institutional mechanisms.

These experiences, where Africans have developed mechanisms to speak with a unified voice in international forums, represent pre-eminent examples of the operational realization of Pan-Africanism.[5] The expression of Pan-African solidarity should be called for with greater urgency as Africa is facing major challenges from climate change. Therefore, Africa must contextualize the urgency of addressing climate change in international climate negotiations within the framework of its objective to achieve an African Renaissance and Pan-Africanism. African heads

2 Mora, C., Frazier, A., Longman, R. et al., "The projected timing of climate departure from recent variability". *Nature* (2013): 183–187.
3 deMenocal, P., and Tierney, J. "Green Sahara: African Humid Periods Paced by Earth's Orbital Changes". *Nature Education Knowledge* 3(10) (2012): 12.
4 Bernstein, S. "Rio+ 20: Sustainable development in a time of multilateral decline". *Global Environmental Politics* 13(4) (2013): 12–21.
5 Sentime, K. "Pan-African Unity as a Pre-Requisite for Pro-Active Response to Climate Change". In Oghenerobor Benjamin Akpor, Phindile Lukhele-Olorunju, and Mammo Muchie (eds.), *The African Union Ten Years after: Solving African Problems with Pan-Africanism and the African Renaissance.* Pretoria: Africa Institute of South Africa, 2013,: 343–364.

of state have established the CAHOSCC coordination mechanism, which aims to improve the prospects of members' expressing more easily and solidly their solidarity on environmental issues and climate change, most notably during the January 2013 African Union (AU) summit in Addis Ababa. The temporary merger of the rotating body presidency of the African Ministerial Conference on the Environment (AMCEN), which is a conference structure that regularly brings together African Ministers of the Environment, with the coordinating power of the CAHOSCC enables the timelier and more synergistic articulation of Africa's interests on the continent and global levels. This rationalization has improved the prospects for promoting consensus and facilitating unity between member states on cross-cutting links characterized by the impacts of climate change on various sectors.

Climate Change Challenges and Prospects for African States

The CAHOSCC coordination mechanism is a platform that potentially catalyzes Pan-African solidarity and efficiency in advancing the environmental interests of Africa in the cause of addressing climate change.[6] Only a decisive paradigm shift in conceptualizing climate change taking place nationally, continentally, and globally could help to improve prospects for addressing this challenge. The massive accumulation and concentration of greenhouse gases endangers human survival, and historical emitters of carbon emissions must take steps to reduce their own emissions while also investing in emission control programs in Africa and other developing regions, as mitigation measures cannot be considered in isolation.[7]

At the national level, awareness of the urgency to act against climate change has led states to adapt national policies to combat these problems. For example, the United Nations Framework Convention on Climate Change (UNFCCC) state parties, including West African states, have adopted National Adaptation Policies (NAPs), often implemented through mechanisms such as Intended Nationally Determined Contributions as tools for implementation and follow-up of defined pol-

---

6 Deressa, T., Hassan, R., and Ringler, C. "Perception of and Adaptation to Climate Change by Farmers in the Nile Basin of Ethiopia". *The Journal of Agricultural Science* 149 (1) (2011): 23–31.
7 Tadesse, D. "The impact of climate change in Africa". *Institute for Security Studies Papers* 220 (2010): 1–20.

icies and provisions of the UNFCCC. However, addressing climate change as a global issue remains stifled by the predominance of national or categorical interests.[8]

The intersection of national interests, global finance corporations, and greenhouse gases emitters makes it possible to understand the minimum number of agreements international climate change negotiations. For example, the 2015 Paris Conference of the Parties (COP 21)[9], revealed the fragile position of African states during climate negotiations. Despite the initiatives taken at the continental level and the need to define a common African position, states are individually represented, and neither the AU nor other African regional organizations are parties to the UNFCCC. Taken individually, they are unable to properly defend their interests, which aggravates their weaknesses in terms of expertise, political, and economic weight. International and continental networks of civil society organizations and local authorities paradoxically seem to replace states in defending a common African position on climate change.

The fight against climate change is also affected by weak governance. The diverse causes of weak governance and its political, economic, social, and environmental causes demonstrate that climate change should be treated as a governance issue.[10] There are two reasons for this. First, climate change as a phenomenon exhibits an entanglement of causes, consequences, and links between diverse sectors such as energy, water, forests, agriculture, transport, urbanization, etc. Second, climate change independently reveals its effects on all sectors in total interdependence, from local to global. Considerable efforts have been made to promote a holistic approach more adapted to the problem.[11] However, this problem is compounded by the inadequacies of global governance, as the precedence of national sovereignty coupled with the absence of legally binding obligations becomes a detriment to building a true community of interest around addressing climate change. At the national level, mitigation and adaptation objectives are transversally marked by intrinsic, public action to address climate change, yet they are characterized by fragmented policies and compartmentalization of actors.[12]

---

8 Raleigh, C. "Political marginalization, climate change, and conflict in African Sahel states". *International studies review* 12(1) (2010): 69–86.
9 Robbins, A. "How to understand the results of the climate change summit: Conference of Parties21 (COP21) Paris 2015". *Journal of Public Health Policy* 37 (2016): 129–132.
10 Raleigh, "Political marginalization", 69–86.
11 Butt, T., Bruce, M., Angerer, J. et al. "The economic and food security implications of climate change in Mali". *Climatic change* 68(3) (2005): 355–378.
12 Nyong, A., and Niang-Diop, I. "Impacts of climate change in the tropics: the African experience". In Hans Joachim Schellnhuber, Wolfgang Cramer et al. (eds.), *Avoiding dangerous climate change*. Cambridge: Cambridge University Press, 2006, 235–243.

National policies are still very sectoral and guided by the institutional logic of each field and the habits of the actors responsible for their definition and implementation. Stakeholder interventions suffer from a real coordination deficit, harmonization, and task allocation, which impacts their effectiveness and efficiency in allocating resources. Local authorities are among the first victims of climate change because they suffer from its direct effects. However, Article 35 of the Niger Constitution states that "Everyone has the right to a healthy environment." The state has an obligation to protect the environment in the interest of present and future generations."[13] Everyone is obliged to contribute to safeguarding and improving the environment where they live. Acquisition, storage, handling, and disposal of toxic or polluting waste from factories or other industrial or craft units located in the national territory are regulated by law.[14]

Transit, import, storage, landfill, discharge of toxic waste or foreign pollutants into the national territory, as well as any agreement relating thereto, are constituted as a crime against the nation and punishable by law. The state ensures the evaluation and control of the impacts of any project and program of development on the environment. For instance, Article 37 of the Niger Constitution states that "national enterprises and international organizations have the obligation to respect the legislation in force." They are required to protect human health and to contribute to safeguarding and improving the environment. "

The Niger Prime Minister's General Policy Statement dated June 16, 2012 specifies that regarding the sustainable management of natural resources and the protection of the environment, "the focus will be on the sustainable management of ecosystems, through actions to combat land degradation and sand dunes, the realization of the Great Green Wall and efforts to adapt to changes in climate change to ensure the sustainability of the productive base of agriculture."[15] The Regional Center for Training and Applications in Agrometeorology and Operational Hydrology (AGRHYMET), the African Centre of Meteorological Applications for Development (ACMAD), the Research Institute for Development (IRD), and bilateral and multilateral cooperation organizations were established in addition to these international training and research institutions. To coordinate reflection and action on

---

13 "Constitution of the fifth Republic of Niger July 18, 1999". *International Labour Organization*, last modified April 20, 2022. Available online at: https://www.ilo.org/dyn/natlex/docs/ELECTRONIC/61825/90477/F1990161536/NER61825.pdf

14 Sissoko, K., van Keulen, H., Verhagen, J. et al. "Agriculture, livelihoods and climate change in the West African Sahel". *Regional Environmental Change* 11(1) (2011): 119–125.

15 O'Connor, D., and Ford, J. "Increasing the effectiveness of the "Great Green Wall" as an adaptation to the effects of climate change and desertification in the Sahel". *Sustainability* 6(10) (2014): 7142–7154.

the major issues in accordance with the Rio agreements on the environment and development and Agenda 21, Niger established the National Council on Environment for Sustainable Development (CNEDD) in January 1996, which is the body responsible for the coordination and monitoring of the national environmental and health policy sustainable development. The CNEDD has drawn up the National Environment Plan for Sustainable Development (PNEDD), which serves as a framework of inspiration for all policies in the field of the environment and sustainable development. The Plan is composed of six priority programs: the National Action Program to Combat Desertification and Natural Resources Management, the Climate Change Program, the National Strategy and Action Plan on Biodiversity, and others.[16] The CNEDD is the national body for coordinating and monitoring the related activities according to the post-Rio conventions and their protocols and any other conventions that Niger would subscribe to. As a result, it ensures coordination and follow-up of interventions related to the PNEDD and all other post-conference activities of the United Nations Conference on Environment and Development. The National Technical Committee on Climate Change and Variability (CNCVC) was set up in July 1997. On the legal front, Niger has signed and ratified several conventions and agreements in the field of the environment: three post-Rio conventions (the Convention on Climate Change, the Convention on the Fight against Desertification, and the Convention on Biological Diversity) and the Kyoto Protocol.

These steps automatically resulted in more expensive coping mechanisms for the Sahel, Obviously, the mitigating factors for historical transmitters and adaptation demands for the region can't be considered in isolation, but global sensitivity is needed to deal with climate crises. Climate change will reduce ecosystem tolerance and capabilities for adaptation to biotic systems.[17] Economic policies not rooted in the service sphere may not improve the resilience of institutions and utilization of natural resources. Additionally, knowledge gaps tend to weaken the effectiveness of decision-makers, which translates into sub-optimal scientific, economic, social, and sectoral dimensions for addressing climate change.[18] Legal and

---

16 Odada, E. O., Lekan, O., and Oguntola, A. J. "Lake Chad: experience and lessons learned brief". In "Experiences and lessons learned: Brief for Lake Chad 2005", *Semantic Scholar*, last modified April 21, 2022. Available online at: https://www.semanticscholar.org/paper/Lake-Chad%3A-experience-and-lessons-learned-brief-Odada-Oyebande/c081a17d3bf32351813a870343399d6277e80dc0#citing-papers
17 Desanker, P., and Justice, C. "Africa and global climate change: critical issues and suggestions for further research and integrated assessment modeling". *Climate Research* 17(2) (2001): 93–103.
18 Butt, "The economic and food security", 355–378.

institutional systems have failed to reverse the patterns of natural resource consumption that reinforce the constraints posed by climate change.

Given the previous risks, it would be logical to affirm that Africa's potential to adapt would be severely challenged, unless the quest for sustainable development takes center stage in efforts to achieve significant economic change.[19] Considering the comparative and historical dimensions of regional and national variations in gaseous pollution based on a fair and equitable analysis, the global challenge of climate change should be addressed based on historical grounds and the principle of differentiated responsibilities.

Although there is international awareness of climate change issues, the mechanisms to address them remain largely national, due to the lack of multilateral regulatory mechanisms and environmental laws. The absence of multilateral mechanisms not only reduces the opportunities for effective long-term action but also pushes states to unilaterally manage environmental issues. This is particularly true with respect to security impacts. In essence, security has a strong national dimension.

Effective multilateralism is crucial to address security issues in the Sahel, with their attendant military, human, national, and international dimensions. In this regard, the primary concerns of the Sahel states, which are linked to climate change, human security and well-being, and food security, are at the heart of their national development strategies. Effective policy responses to the climate-change security nexus go through broad development strategies and coordination of various areas of national policy, which go beyond purely security responses.

The Role of Institutions in the Sahel Region

Since 1992 the international community has been trying to deal with climate change through the UNFCCC.[20] Several institutions have been set up to respond to the political, technical, and technological dimensions of climate change and the need to take up responsibility for them from the perspective of global solidarity. These include the Conference of the Parties (COP), the Subsidiary Body for Scientific and Technological Advice (SBSTA), and constituted bodies such as the Executive Board of the Clean Development Mechanism, the Advisory Expert Group, the

---

19 Nyong, "Impacts of climate change", 237.
20 Agnew, C., and Warren, A. "A framework for tackling drought and land degradation". *Journal of Arid Environments* 33(3) (1996): 309–320.

Expert Group on Technology Transfer, the Least Developed Countries Expert Group, and Party Groupings (Island States, the European Union, etc.).[21]

Financing mechanisms are also used to enhance resilience to climate change, limit the emission of greenhouse gases, and promote the transition towards sustainable development. At the continental level, there are also spaces for consultation on climate change problems. In Africa, the AU without being a party to the UNFCCC attempts to develop a common and shared vision on climate challenges which the continent is facing. It relies on the Committee of African Heads of State and Government on Climate Change (CAHOSCC), the African Ministerial Conference on the Environment, the African Ministers' Council on Water (AMCOW), the African Group of Negotiators on Climate Change (AGN) and executives, like ClimDev-Africa.[22] Likewise, regional organizations, such as the Economic Community of West African States (ECOWAS), have encouraged intergovernmental dialogue for disaster risk management related to climate change. Within the West African Economic and Monetary Union (WAEMU), a common policy to improve the environment was adopted in 2012, although it does not grant a high priority to climate change risk management.[23] Some concerted initiatives on climate change exist in the Permanent Interstate Committee for Drought Control in the Sahel (CILSS) and G5.

The Sahel regional states' priorities are to implement change-related programs and projects on climate change and to achieve development goals, including the Millennium Development and Sustainable Development Goals to combat poverty, by focusing on achieving food security, especially for the most vulnerable groups. In the context of multilateralism, developed countries and partners must fully support the implementation of adaptation strategies and plans in the Sahel.[24] In preparing the climate change strategy for the Sahel, it is important to recognize that adaptation is a top priority for the region. Hence, there is an urgent need to implement adaptation measures through the provision of new and additional public financial resources, sound technology, and capacity building in a predictable and

21 Thomalla, F., Downing, T., Spanger-Siegfried, E. et al. "Reducing hazard vulnerability: towards a common approach between disaster risk reduction and climate adaptation". *Disasters* 30(1) (2006): 39–48.

22 Wilby, R. Climate for development in Africa (ClimDev) – climate sciences and services for Africa. Strategic research opportunities for ClimDev-Africa. Report (Loughborough: Loughborough University, 2014). *Handle.Net*, last modified April 20, 2022. Available online at: https://hdl.handle.net/2134/16269

23 Epule, T., Ford, J., Lwasa, S., and Lepage, L. "Climate change adaptation in the Sahel". *Environmental Science & Policy* 75 (2017): 121–137.

24 Conway, D. "Adapting climate research for development in Africa". *Wiley Interdisciplinary Reviews: Climate Change* 2(3) (2011): 428–450.

timely manner. The goal of adaptation is to move from vulnerability assessment to the implementation of adaptation programs. A significantly improved assessment of adaptation costs for developing countries, especially in Africa, is also necessary.

The analysis of the implications of technology, finance, and capacity building in different types of adaptation programs must distinguish between short and long-term adaptation measures, autonomous adaptation (and therefore complementary to development planning) and those integrated into development planning. It should include the establishment of a network of African centers of excellence on climate change, and a regional risk information system for short, medium, and long-term climate change in Africa. Adaptation must be dealt with in a consistent manner under the UNFCCC, and the implementation of effective adaptation strategies needs to be promoted.

If the impacts of climate change are not defined with greater certainty, adaptation policies will have to focus on reducing the vulnerability of populations to climate variability, a major feature of the Sahelian climate. Vulnerability to climate variability, namely rainfall, is high in the event of a highly rain-dependent agricultural sector. Agricultural development and food security are development priorities in the Sahel. There is a natural overlap between climate change adaptation strategies and policies to increase food production and resilience to climate variability. Given the severity of the projected impacts of climate change, climate change adaptation has become a new priority for the development community. It is undeniable that "good adaptation" to climate change is synonymous with "good development." This is the reason for the identification of vulnerabilities and long-term planning in policymaking. But this sense of urgency is not without risk, as the need to "do something" in the face of anthropogenic climate change should not override other major development aims.

For the Sahel, the impacts of climate change are a development issue, and investment in development is the best instrument for promoting peace and security. This interpretation finds its counterpart in the concept of human security, which encompasses concerns such as well-being, food security, and environmental safety. To integrate the long-term features of climate change into national and regional development strategies, the effects of all the drivers of change in the Sahel and their interactions with climate change must be analyzed. Demographic dynamics, migration, trade, and economic development are some of the factors to be analyzed.

These long-term dynamics will be crucial for understanding and sustainably addressing climate change vulnerability. Promoting orientation on climate change and its effects at the level of regional African institutions, such as the AU, ECOWAS, or the Intergovernmental Authority on Development (IGAD) should also be a priority. Identifying regional concerns and strategies improves the coordination and efficiency of activities. International partners should support efforts to formulate regional action plans and policy responses to climate change and engage in dialogue based on identified priorities. During the process, consideration should be given to integrating climate issues into their regional strategies and their links to regional instability. Early warning mechanisms are an important tool for conflict prevention. There are various early warning and monitoring mechanisms in Africa, often within regional or continental intergovernmental forums. Most are designed and developed on a purely military basis.

Integrating environmental variables into the monitoring and analysis of early warning mechanisms would widen the range of insecurity signals. Some data on seasonal climate forecasts and food security dynamics are already available at the regional level. As a first step, centers specializing in climate science, early warning mechanisms for conflict prevention, and partner institutions could be brought closer together. It would initiate a dialogue process to define intra-African methodology, variables to be included, sources of data, and data exchange and share best practices. European partners could support this dialogue process and provide additional data in addition to technical and financial assistance. Preliminary informal discussions with African stakeholders and Organization for Economic Cooperation and Development (OECD) partners have confirmed the interest in this proposal.

Security Challenges of Water Scarcity in the Sahel Region

Water scarcity and climate change are regularly cited as the most serious crises that humanity will face in the coming decades. In fact, the links between the two issues are so close that they should be seen as a single issue. The effects of climate change on the global physical landscape are changing the geopolitical situation and destabilizing water-scarce regions like the Sahel.[25] It can undermine the ability of countries to govern themselves and can generate unexpected conflicts. Unlike other international security risks, climate change can be modelled with a relatively high degree of certainty. Growing water scarcity in the Sahel is causing

---

25 Misra, A. K. "Climate change and challenges of water and food security". *International Journal of Sustainable Built Environment* 3(1) (2014): 153–165.

water levels in traditional wells to drop, forcing people to travel long distances to collect limited amounts of water.[26]

Water scarcity in the Sahel is exacerbated by many factors, including climate change, increasing water demand, and population growth. In recent decades, the impact of climate change has manifested itself in the form of increased variability in precipitation, temperature, and wind speed. These have contributed to water scarcity, droughts, floods, sandstorms, and heavy rain. The increase in water scarcity also has wider implications for families' livelihoods and may raise the risk of migration, urbanization, and conflict. Lack of water resources and poor water quality also increase the risk of diarrhea among young children.[27] Repeated bouts of diarrhea in young children can lead to stunting and malnutrition and irreversibly damage their physical and mental development. Women and girls are mainly tasked with collecting water, which exposes them to increased safety risks and increased physical burden. More time spent collecting water also reduces the time and energy available for school, which affects children's education, attendance, and participation, and therefore their future opportunities.[28]

Mali, Niger, and Chad are among the countries crossed by the Sahel's fragile economic belt despite being able to receive as much water as tropical regions, such as northern Cameroon or central Nigeria. This evolution, already visible during increasingly frequent and destructive thunderstorms, is linked to the warming of neighboring oceans.[29] The phenomenon generates stronger evaporation than before, while the monsoon winds from the Atlantic strengthen and shift to the north. Rainfall in turn releases heat, transforming the process into a vicious cycle. In other words, a self-amplification mechanism is formed due to the increase in the Earth's temperature. While crossing this tipping point could be potentially beneficial for the region's climate, the change could be so huge that it would be a major adaptation challenge for an unstable region, particularly for agriculture in this sensitive and poor area.[30] The rains become more violent, flooding, and gullying the soil, ravaging crops.

26 Cooper, R., and Price, R. "Unmet needs and opportunities for climate change adaptation and mitigation in the G5 Sahel region. K4D Emerging Issues Report". Brighton, UK: Institute of Development Studies, 2019, 1–51.
27 Sartori, N., and Fattibene, D. "Human Security and Climate Change. Vulnerabilities in the Sahel". *Euromesco Policy Brief 94* (2019): 1–10.
28 Niasse, M., and Varis, O. "Quenching the thirst of rapidly growing and water-insecure cities in sub-Saharan Africa". *International Journal of Water Resources Development* 36 (2–3) (2020): 505–527.
29 Keys, P. et al. "Invisible water security: Moisture recycling and water resilience". *Water Security* 8 (2019): 100046.
30 Anderson, Z. K. *Water Scarcity and Violent Conflict in Nigeria.* Monterey: Naval Postgraduate School, 2019.

Besides conflicts, extremism, climate change, and poverty, 150 million Sahelians face immense challenges in water access. In addition, there is a great possibility of a demographic explosion, as the population of the region is set to double in the next twenty years. The Sahel has dealt with chronic climate change for several decades, and frequent floods threaten the survival of a largely agrarian population.[31] 98% of water for agriculture comes from rain, and with more frequent climate shocks, vulnerable households are less able to cope with crises and struggle to recover. Water is becoming increasingly scarce in Sahelian countries, which often suffer from a structural deficit of water infrastructure.

Lack of access to water has been exacerbated by conflict or insecurity in some regions. The 2012 Mali Conflict led to the flight of all technical state services in the north, resulting in a significant reduction of places with water access, which were destroyed, pillaged, or simply left abandoned.[32] Some villages are sorely lacking in this vital resource, which turns water scarcity into a migration issue. People move to find areas where there is water to drink, wash, irrigate crops, or water their livestock. Northern Cameroon, historically lacking access to basic services, has faced the displacement of massive populations caused by insecurity and violence. Due to 200,000 people displaced in 2016, populations in some host villages have increased dramatically, creating additional pressure on already scarce water resources and the risk of inter-community conflicts.[33]

During the past 20 years, water availability in the region has dropped by 40% and has also become a major health issue. Lack of adequate water, sanitation, and poor hygiene conditions are exacerbated by limited or even failing health systems. These problems are obstacles to adequate healthcare and make some communities extremely vulnerable to waterborne diseases and epidemics. The lack of access to drinking water is also an aggravating factor in malnutrition, a longstanding scourge of the Sahel which continues to reach critical levels in certain regions.[34]

Severe acute malnutrition exceeds the emergency threshold in half of Chad. While the availability of water per capita has decreased by more than 40% in

31 Mapedza, E., Tsegai, D., Bruntrup, M., and McLeman, R., (eds.), *Drought Challenges: Policy Options for Developing Countries.* Amsterdam: Elsevier, 2019.

32 Adaawen, S., Rademacher-Schulz, C., Schraven, B., and Segadlo, N. "Drought, migration, and conflict in sub-Saharan Africa: what are the links and policy options?" *Current Directions in Water Scarcity Research* 2 (2019): 15–31.

33 Ruppel, O. C., and Funteh, M. B. "Climate change, human security and the humanitarian crisis in the Lake Chad Basin region: selected legal and developmental aspects with a special focus on water governance". *Law| Environment| Africa* 1 (2019): 105–136.

34 Falkenmark, M. "Water resilience and human life support-global outlook for the next half century". *International Journal of Water Resources Development* 36(2–3) (2019): 377–396.

the last twenty years in Sahelian countries, joint efforts by humanitarian action, development actors, and governments should continue to provide Africans with sufficient and lasting access to this precious resource. Further, increasing water scarcity leads to political unpredictability and conflicts. Several other scenarios show how the interweaving of climate effects and the security situation result in the formation of a new geopolitical landscape. Fast-growing coastal cities in the Horn of Africa, such as Mogadishu, Djibouti, and Mombasa, are vulnerable to rising sea levels.[35] The sea could flood critical urban infrastructure, contaminate freshwater resources with the intrusion of saltwater, and reduce arable land, driving the population into migration. The Gulf of Aden is a crucial maritime basin in the Sahel region. Climate change is reducing the region's meager economic opportunities, tending to increase piracy along the coast.[36]

However, there is a significant overlap between countries with the highest climate vulnerability in Africa and those with a high incidence of pirate attacks, such as Somalia and Eritrea. This disturbing overlap of risks can prolong the failure of Horn of Africa states. The militarization of water supplies and changes in the availability of water resources have enabled states and non-state actors to use water as a weapon. According to recent research by Marcus King from George Washington University, Somalia is particularly exposed to this conjunction between climate, conflict, and the militarization of water resources.[37] Regional droughts experienced by Somalia in 2011 have been linked to climate change. The fundamentalist jihadist group Al-Shabab changed its guerrilla tactics and began to attack cities to project power and establish a presence. Climate change, food shortages, and the continuing conflict with the militarization of water supplies, coupled with the difficulty of accessing humanitarian aid, have had serious consequences for the population. It has resulted in more than 250,000 deaths and hundreds of thousands of displaced people.[38]

Climate disturbances around the world have a pronounced impact on the water resources available in the Sahel region. In recent years, this semi-arid tran-

---

35 Agutu, N. O., et al. "GRACE-derived groundwater changes over Greater Horn of Africa: Temporal variability and the potential for irrigated agriculture". *Science of the Total Environment* 693 (2019): 133467.
36 Adaawen, Rademacher-Schulz, Schraven, and Segadlo, "Drought, migration", 15–31.
37 "Epicenters of Climate and Security, June 2017: the new geostrategic Landscape of the Anthropocene", *The Center for Climate and Security*, 2017, last modified April 25, 2022. Available online at: https://climateandsecurity.files.wordpress.com/2017/06/epicenters-of-climate-and-security_the-new-geostrategic-landscape-of-the-anthropocene_2017_06_091.pdf
38 King, M., and Burnell, J. "The Weaponization of Water in a Changing Climate". In Caitlin E. Werrell and Francesco Femia (eds.), *Epicenters of Climate and Security: The New Geostrategic Landscape of the Anthropocene*. Washington: Center for Climate and Security, 2017, 67–73.

sitional ecoregion has faced serious and enduring problems, including the adverse effects of climate change, irregular rainfall, and recurrent droughts that have limited crops. Groundwater is the main source of water for many people. Aquifer withdrawals are increasing, but not properly regulated. It has resulted in the overexploitation of water resources, whose quantity and quality have decreased.[39] As surface water supplies are limited, the countries of the Sahel draw for their consumption of drinking water from the underground waters of one of the five aquifers: the Iullemeden Aquifer System, the Liptako-Gourma-Upper Volta system, the Senegalo-Mauritanian Basin, Lake Chad, and Taoudeni basins.

These transboundary groundwater resources are shared by thirteen African International Atomic Energy Agency (IAEA) member states: Algeria, Benin, Burkina Faso, Cameroon, Ghana, Mali, Mauritania, Niger, Nigeria, Central African Republic, Senegal, Chad, and Togo. In the context of Africa, water is discussed in terms of need or abundance. An example of this is a remote village in Ghana that is almost depopulated, as only saltwater flows from its well.[40] The difficulty in obtaining drinking water eventually forced most of its residents to abandon it and relocate closer to a safe water source. A limited understanding of how aquifers work and the lack of guiding principles or rules for the use of groundwater in most countries of the Sahel can lead to over-exploitation, pollution, and degradation of these resources.

Due to a lack of knowledge, wells can be drilled and then immediately abandoned because the water supply is insufficient. In the Sahel, fetching water has traditionally been the responsibility of women and young children, and women sometimes must travel miles before reaching the nearest source of drinking water. As more people use underground aquifers as their main source of drinking water, concerns arise: what is the volume of these groundwater resources? Can we count on it to feed the Sahel region in the future? In 2012, the IAEA launched a large-scale, four-year technical cooperation project to promote integrated management and the development of shared groundwater resources in the Sahel region.

This project aims to map groundwater through the application of isotopic techniques in hydrological studies and to identify and understand the root causes of the main threats to the five transboundary aquifers. Isotopic hydrology techniques can also provide useful information on the quality and availability of hidden

---

39 Niyitunga, E. "A conceptual analysis for understanding water scarcity and its threats to international peace and security". *African Journal of Public Affairs* 11(3) (2019): 95–117.
40 Kwame, A.-M., and Kusimi, J. M. "Dwindling water supply and its socio-economic impact in Sekyere Kumawu District in Ashanti Region of Ghana: public opinion on the role of climate change". *GeoJournal* 85 (5) (2020): 1355–1372.

groundwater and analyze the impact of climate change on water resources.[41] By following the path of isotopes in water, scientists can obtain valuable information quickly and inexpensively, and thus have a better understanding of water resource systems. This isotopic data can help countries shape improved water management strategies and climate change adaptation policies to sustainably meet their current and future water needs. As demand grows in the face of limited water resources, the question of managing aquifers shared between several countries is becoming more pressing.[42] The Sahel countries recognize the importance of cooperating to put in place, in an integrated manner, the technical, legal, and institutional frameworks necessary for the management of these resources.

As a part of its technical cooperation program, the IAEA is now helping 13 African countries in the Sahel to monitor and determine the characteristics of aquifers using isotopes, so they can better understand how transboundary aquifer systems work, how much water each country can extract without drawing on the reserves of another, and what impact human activities have on aquifers. This information is essential for the development of effective regional water management programs. Scientists from the Ghana Atomic Energy Agency have set up a laboratory for the research of tritium with support from the IAEA's technical cooperation program. Ghana is one of the countries in the region to have acquired modern equipment through the Sahel project. Technicians are in training, and the laboratory is being expected to play a central role in data analysis within the region. The mapping and knowledge of precious water resources will help ensure that countries in the Sahel region can develop long-term strategies for equitable, sustainable sharing and management of freshwater resources.

The Sahel is one of the regions most affected by global climate inequalities. Regional states are responsible for a tiny share of global greenhouse gas emissions, estimated at 0.25% with the per capita emissions of four countries (Burkina Faso, Chad, Mali, and Niger), which are among the lowest 10% in the world. According to the GAIN Index, the Sahelian countries of West Africa rank among the 20% of the most vulnerable countries in the world, with Niger being the most vulnerable country, and Chad being third place. Climate change is already at work in the Sahel, which has become drier and warmer. The temperature has risen faster than the world average, and average precipitation has decreased in all countries. The poor distribution of the rain in time and space causes episodes of drought fol-

---

41 Eddi, M. *Cirad-Highlights 2018: Activities report.* Paris: CIRAD, 2019.
42 Döring, S. "Come rain, or come wells: How access to groundwater affects communal violence". *Political Geography* 76 (2020): 102073.

lowed by floods, destroying crops during floods or causing large population displacements.

For example, the 2009 Ouagadougou floods forced 150,000 people to flee their homes. As for water problems, since the 1970s and 1980s, emphasis has been placed on the water supply to rural populations and village irrigation systems. Within the broader framework of decentralization policies, territorial communities have gradually become frontline players in the provision of these services. Today, small-scale irrigation still accounts for most irrigation in the area, but communities have also involved private operators for construction, as well as management and equipment maintenance. In the 1990s and 2000s, private irrigation significantly contributed to the expansion of irrigated areas in the Sahel. Despite these changes, the areas suitable for irrigated agriculture in the Sahel remain underdeveloped and not fully exploited.

Sahel countries still have many untapped water resources. Failing to use reserves of surface and underground water, agriculture remains mainly rainfall dependent and directly subjected to climatic hazards. The question of water access arises as much for humans as it does for livestock, especially in areas where pastoral farming is the primary economic activity. This situation fully justifies the priority given to improving access to water in the PIP Emergency Program. The different targeted infrastructures will improve the living conditions of rural populations in these border areas by reducing waterborne infectious diseases and allowing the development of production irrigated and easier access to water for pastoral populations. These actions will demonstrate to the population that governments are involved in satisfying their basic needs and do not just provide a security response.

The huge human, natural, and cultural resources give the Sahel considerable potential for growth. However, this region also faces deep-seated environmental, political, and security challenges that undermine prosperity and peace. The combined effect of poor livelihoods, natural disasters, conflict, and migration has increased competition for access to scarce natural resources. This is particularly true for the agriculture sector, on which two-thirds of the population depends. Pastoral land and water form the backbone of Sahelian economies and societies. Population growth, urbanization, and changes in consumption patterns are all the factors allowing the agricultural sector to experience constant change, which opens the way to new opportunities for the transformation of agriculture and Sahelian food systems. However, the challenges mentioned above pose a threat to the economy, food security, and vital ecosystems of the entire region. In recent decades, the United Nations (UN) has acquired solid experience and a comparative advantage in the Sahel. Thus, the engagement of the UN as a partner provides access to this unique combination of skills and expertise. While this chapter tends to invalidate a direct causal relationship between climate change, migration, and conflict in the region, the combination of livelihood vulnerabilities, exacerbated by changing climatic conditions and non-climatic factors, can lead to behavioral responses such as conflict and migration.

Tatiana Konrad

Chapter IV
Violence in the Age of Environmental Crisis: Climate Change Denialism, the War on Science, Eco-Anxiety, and *Don't Look Up*

"You can't blame the people who can't hear the warnings; you *have* to blame the ones who can, and who ignore them."[1]

"People are actually afraid to be interested [in environmental conversation], because they suspect . . . that we'll find if we dig deep enough that we've gone so far beyond the limits of what the planet will tolerate that only a major catastrophe which cuts back both our population and our ability to interfere with the natural biocycle would offer a chance of survival."[2]

The quotations that open this essay are from John Brunner's novel *The Sheep Look Up* (1972). Directly responding to environmental concerns of that time – the use of pesticides and other chemical fertilizers to increase food production in the post-World War II times, the napalm attacks and their impact on humans, nonhumans, and the environment during the Vietnam War but also prior to it, to name just a few – the novel imagines an apocalyptic world as a result of environmental degradation as well as raises the problem of humanity's action, or lack thereof, to restore environmental health and thus save the planet. The novel emphasizes that the lack of information is not what causes further inaction – after all, people can see and experience environmental degradation. Yet most of the characters in the novel continue to neglect the situation and accept the inevitable environmental collapse as the only possible future. Why would they do that? Is not a future something that one would want to have? As past and present choices and preferences of certain individuals, groups of people, and even nations demonstrate, including the use of fossil fuels and dependence on cheap energy, humanity is more interested in temporary comforts, and any potential change that might deprive one of these comforts is not welcome. This has much to do with human psychology and behavior. *The Sheep Look Up* imagines "a major catastrophe"[3] that takes responsibility away from humans and restores some kind of balance.

---

1 John Brunner, 1972, *The Sheep Look Up*. New York: Open Road Integrated Media, 2016, 228. [Italics in original.]
2 Brunner, *The Sheep Look Up*, 350.
3 Brunner, *The Sheep Look Up*, 350.

https://doi.org/10.1515/9783111081687-005. The original version of this chapter has been revised. Unfortunately, the name of the contributor was misspelled in the original publication due to a production error. This has been corrected, along with some typographical errors. The press apologizes for any inconvenience caused.

Today, half-a-century later, the black comedy *Don't Look Up* (2021) raises very similar issues – why would the majority neglect a catastrophe that would destroy the planet? – and provides almost identical answers – because someone would find a way to save the planet, and if not, then so be it!

This chapter analyzes the catastrophic scenario outlined in *Don't Look Up* and views it as a metaphor for the ongoing environmental crisis that humanity faces today. The essay focuses on the lack of sufficient action that could help save the planet. It explores climate change denialism and the war on science as forms of violence. Drawing on Rob Nixon's important concept of "slow violence,"[4] the essay emphasizes the necessity to re-envision violence today, in the era of environmental crisis, and explores the danger of invisible, non-immediate forms of violence that lead to such serious consequences as environmental degradation and environmental inequality. The present chapter draws on *Don't Look Up* to illustrate the danger of climate change denialism and the war on science, as depicted in the film. It seeks to answer the following questions: Why does humanity refuse to believe in the end of the world? Why cannot humans change and give up on the idea of living in the world-as-we-know-it, choosing only pro-environmental ways of living and co-existing? Why, despite the ongoing environmental degradation, do some individuals continue to deny the existence of environmental problems? And finally, can one interpret such behavior as a form of violence?

Violence: Climate Change Denialism and the War on Science

Violence is often viewed as a form of physical or psychological pressure exercised by one party over the other. One can see violence and its effects when they are quick or even immediate. This is true for some forms of violence, but not for all. Drawing on the environmental humanities and postcolonial studies, Rob Nixon identifies another form of violence, "slow violence," that he defines as "a violence that occurs gradually and out of sight, a violence of delayed destruction that is dispersed across time and space, an attritional violence that is typically not viewed as violence at all."[5] Nixon foregrounds that traditional, or more common, understandings of violence are very limited and limiting in today's world:

---

4 Rob Nixon, *Slow Violence and the Environmentalism of the Poor.* Cambridge: Harvard University Press, 2011.
5 Nixon, *Slow Violence*, 2.

Violence is customarily conceived as an event or action that is immediate in time, explosive and spectacular in space, and as erupting into instant sensational visibility. We need . . . to engage a different kind of violence, a violence that is neither spectacular nor instantaneous, but rather incremental and accretive, its calamitous repercussions playing out across a range of temporal scales. In so doing, we also need to engage the representational, narrative, and strategic challenges posed by the relative invisibility of slow violence.[6]

Nixon largely associates "slow violence" with the postcolonial world, and the instances of environmental racism and inequality that postcolonial nations have been experiencing vividly illustrate the brutal effects of "slow violence." Yet, violence as such, and "slow violence" in particular, is much more complex and goes beyond the postcolonial world. One way violence is experienced globally today is via climate change denialism and the war on science. Sabotaging the ongoing attempts to minimize the effects of climate change, stop environmental decline, and thus prevent a global environmental catastrophe, climate change denialism and the war on science are forms of violence that allow and support the exploitation of the environment, the destruction of the nonhuman world, and the environmental inequality experienced by certain individuals and nations. Climate change denialism and the war on science are political instruments that have a profound impact on the current environmental choices and decisions made by individuals worldwide and, in a violent way, shape the future that will be informed by this type of violence and be violent in itself.

Environmental humanities scholars persistently emphasize the detrimental role that such forms of violence play today. Nicole Seymour, for example, identifies "[c]limate change denialism/skepticism" as a "problem of environmental/scientific knowledge."[7] Specifically, climate change denialism and the war on science attempt to reshape environmental knowledge in a way that would allow us to temporarily preserve the world-as-we-know-it without resetting priorities, changing practices and behaviors, and making investments that would help preserve a healthy environment. According to Michael E. Mann and Tom Toles, "The climate change-denial industry has adopted the strategy of obscuring the basic concepts through a torrent of typically misleading arguments about technical details and minutia. But all of that never has and never will change the basic fact that more CO_2 in the atmosphere traps more heat and warms Earth's surface."[8] The war on science as such

---

6 Nixon, *Slow Violence*, 2.

7 Nicole Seymour, *Bad Environmentalism: Irony and Irreverence in the Ecological Age*. Minneapolis: University of Minnesota Press, 2018, 41.

8 Michael E. Mann and Tom Toles, 2016, *The Madhouse Effect: How Climate Change Denial Is Threatening Our Planet, Destroying Our Politics, and Driving Us Crazy*. New York: Columba University Press, 2018, 15.

has not emerged in the context of climate change,[9] yet it has been used to manipulate audiences and promote certain political and sociocultural views. One prominent example here is Donald Trump, who has been vigorously denying climate change for years, including before the 2016 presidential election.[10] Today, climate change denialism continues to manifest itself through various speculations that undermine the realness of climate change.[11]

There is a tight connection between violence and catastrophe in the current discourses on climate change. Violence manifests itself through climate change denial, the war on science, and the unwillingness to adjust and make pro-environmental choices, to name but a few. Catastrophe, in turn, has become a descriptive term to emphasize the destructive nature of humanity's anti-environmental actions as well as the events that these actions have already led to or will lead to in the future. Kathryn Yusoff claims that "[t]he catastrophe of climate change is excessive and will inscribe all earthly space."[12] Scholars define the present moment as "a new era of catastrophe risk," claiming that the majority of the catastrophes that happen today are natural catastrophes.[13] Even the catastrophes themselves are described as being catastrophic, like, for example, "the catastrophic fires raging across Australia."[14]

While catastrophe helps emphasize the danger of the ongoing environmental crisis, it also contributes to the spread of climate change denialism and skepticism. As Jairus Victor Grove observes, "Apocalypse is a touchy subject, even for those of us in critical traditions prone to question developmentalist and teleological theories." We often respond to the possibility of catastrophe with skepticism. The practiced intervention is to criticize those proposing the possibility of apocalypse with critiques of eschatological thinking or to argue that representations of the end-time stem from cultural malaise or a reactionary romanticism for simpler times."[15] Grove explains that the reason why there is so much skepticism around apocalypses is because the trope has been largely (ab)used by Hollywood to create scenarios

---

9 "The war on science can be traced back more than half a century, beginning with the activities of the tobacco industry in the 1950s" (Mann and Toles, *The Madhouse Effect*, 69).
10 Mann and Toles, *The Madhouse Effect*, 153–154.
11 Mann and Toles, *The Madhouse Effect*, 154.
12 Kathryn Yusoff, "Excess, Catastrophe, and Climate Change". *Environment and Planning D: Society and Space* 27 (2009): 1010.
13 Howard Kunreuther, Erwann Michel-Kerjan, and Nicola Ranger, "Insuring Future Climate Catastrophes". *Climate Change* 118 (2013): 340.
14 Blanche Verlie, *Learning to Live with Climate Change: From Anxiety to Transformation*. London: Routledge, 2021, 1.
15 Jairus Victor Grove, *Savage Ecology: War and Geopolitics at the End of The World*. Durham: Duke University Press, 2019, 236.

that outline multiple ways in which this world can end.[16] These scenarios appear largely improbable, or they outline a kind of apocalypse that is already well known to underprivileged groups of people, like the ones that depict "adventures of mostly white privileged people having to live like most of the rest of the world does on a daily basis: no food security, the risk of being forced from one's home, unpredictable access to basic things like medicine and emergency care, and terrifying people or zombies or robots coming to get you in the dead of night. There is something undeniably precious about this vision of the apocalypse where people with perfect teeth pretend to be terrified at the possibility of killing and preparing their own food."[17] Nevertheless, no matter how one interprets "catastrophe" and its apocalyptic nature in the context of the current environmental crisis, one fact remains true: "our planet is suffering catastrophic damage from human activities."[18] And the true apocalypse that humanity, and particularly privileged individuals, is afraid of is not the partial or complete destruction of the environment and nonhuman world, not the suffering that will be further inflicted on people of color and other minority groups, but that "a particular way of life" will end – "a way of life that is as threatened by peak oil or any of the other shortages of minerals or capital that are necessary for the predictable routines that many Americans and Europeans have grown accustomed to, undoubtedly at the expense of the rest of the planet's population of human and nonhuman Earthlings."[19]

Don't Look Up draws on the apocalypse to discuss the current environmental crisis. Importantly, the end of the world happens in the film not due to climate change or any other phenomena that have become largely associated with it, including, for example, CO_2 emissions,[20] but via a comet that destroys the planet from space. In doing so, the film draws attention away from climate change and environmental decline and foregrounds the major issue that has impeded environmental progress so far: the inability of the majority, and especially those who have power, to believe that the world can end and that humanity cannot withstand the forces of nature. Through the reference to a comet, the film also reminds the audience that "great extinctions" have already taken place on our planet, and that

16 Grove, *Savage Ecology*, 237.
17 Grove, *Savage Ecology*, 237.
18 E. Ann Kaplan, *Climate Trauma: Foreseeing the Future in Dystopian Film and Fiction*. New Brunswick: Rutgers University Press, 2016, 1.
19 Grove, *Savage Ecology*, 237.
20 For more, see John Urry, *Climate Change and Society*. Cambridge: Polity, 2011, 90. Grove, *Savage Ecology*, 41. Tatiana Prorokova-Konrad (ed.), *Transportation and the Culture of Climate Change: Accelerating Ride to Global Crisis*. Morgantown: West Virginia University Press, 2020.

many of them were caused by "exogenous events such as asteroids."[21] Thus, rejecting the possibility of the end of the world-as-we-know-it is simply wrong. The film illustrates that denialism as a strategy chosen by the characters who have power and could try to save the world is a form of violence that ultimately leads to the destruction of the planet.

Can films like *Don't Look Up* impact audiences and promote pro-environmental behavior and action? *Don't Look Up* is not the first film that plots the end-of-the-world scenario, exploring the issues of power, inequality, and environmental exploitation, among other problems. Films alone certainly cannot change what humans do to the environment. But films visualize environmental crises and invite audiences to see them by changing the perspective: the viewer is given an opportunity to witness the collapse as a distant observer. This helps the viewer to focus on concrete problems that emerge as part of the environmental crisis and thus see the environmental crisis as such. This approach is very valuable, particularly in the Global North, where the effects of climate change are considerably milder compared to the postcolonial world. Films about climate change can spread environmental education. For example, the results of the survey conducted among the viewers of Roland Emmerich's *The Day After Tomorrow* (2004) reveal that the film helped some individuals understand climate change better: "Rather than consider just the typical responses of polar melting and increased storminess, respondents appeared to recognize the multiple dimensions of climate change, including the changing nature of the seasons, and the potential for regional cooling."[22] While some nations are already experiencing the dramatic effects of climate change due to rising sea levels, droughts, and other environmental issues, for many others (these are predominantly privileged nations), climate change remains a distant problem. Such spatial (albeit only temporal) distancing can falsely suggest that one can escape climate change by staying in or moving to a specific place. It also limits the scale of climate change consequences, implying that only those issues that humanity is aware of now are climate change, thus veiling the growing danger of climate crisis that manifests itself in a variety of problems that will continue to grow in the future. Films about climate change make the viewer face climate change. They renegotiate the idea of a global crisis by bringing the crisis into everyone's home.

Such films also reenvision the relationship between collective and personal, both in terms of whom climate change impacts and who is responsible for causing

---

21 Grove, *Savage Ecology*, 235.
22 Thomas Lowe, Katrina Brown, Suraje Dessai, Miguel de França Doria, Kat Haynes, and Katharine Vincent, "Does Tomorrow Ever Come? Disaster Narrative and Public Perceptions of Climate Change". *Public Understanding of Science* 15, no. 4 (206): 452.

climate change and should act to minimize its effects. Climate change films "reframe[e] the perspective from a detached and scientifically-articulated problem to one of a human condition – immediate and personal."[23] *Don't Look Up* negotiates the meaning of climate change on two levels. First, it explicates the relationship between scientists and the general public, emphasizing the importance of bringing science to the public. Second, while zeroing in on individual reactions to the catastrophe (scientists are in a panic, politicians do not care, the rich want to gain financial profit, etc.), the film explores how such reactions inevitably impact and shape collective experiences: humans (and nonhumans) die because of the catastrophe, yet this outcome is the result of personal choices. By denying the catastrophe or giving up in the moment of crisis, the characters make personal decisions that together are a form of collective violence, and that lead to more collective violence at the end of the film.

There is a direct connection between climate change and "collective violence": environmental problems cause sickness, whereas environmental precarity is the reason for migration and even wars.[24] Thus, "climate change undermines human security."[25] Moving beyond anthropocentric visions of climate change, it is crucial to emphasize that climate change and other environmental problems that result from anthropogenic activity negatively impact the more-than-human world, dramatically transforming and destroying it. *Don't Look Up* illuminates the anthropocentric nature of denialism by focusing on human characters only) is a form of anthropogenic collective violence that, in the end, causes violence that goes beyond the human, impacting everything non-anthropogenic too.

Scholars argue that "there is always the dangerous possibility that the more you know about something as potentially catastrophic as climate change the less that you will feel personally responsible for it and the more you may rely on defensive attributions that will shift blame and responsibility elsewhere."[26] *Don't Look Up* explores the consequences of denying, neglecting, and shifting responsibility in the context of environmental crisis. This is effectively illustrated by the general public who, having found out more about the comet and the deadly ram-

23 Ailise Bulfin, "Popular Culture and the 'New Human Condition': Catastrophe Narratives and Climate Change". *Global and Planetary Change* 156 (2017): 141.

24 Barry S. Levy, Victor W. Sidel, and Jonathan A. Patz, "Climate Change and Collective Violence". *Annual Review of Public Health* 38 (2017): 241.

25 Jon Barnett and W. Neil Adger, "Climate Change, Human Security and Violent Conflict". *Political Geography* 26 (2007): 651.

26 Geoffrey Beattie, Laura Sale, and Laura Mcguire, "An Inconvenient Truth? Can a Film Really Affect Psychological Mood and Our Explicit Attitudes towards Climate Change?" *Semiotica* 187, no. 1/4 (2011): 107.

ifications of the catastrophic event, ridicule the catastrophe and the scientists' serious attitude towards it even more.

Don't Look Up is a satire that makes its viewers recognize the criminal nature of denialism and inaction in the era of climate change. It makes the viewers (particularly from Europe and North America) see *themselves* in the characters whom we (the viewers) perceive as stupid, lazy, myopic, and greedy. Robert J. Brulle claims that "the political and cultural arenas" are important "avenues for climate action in the United States."[27] *Don't Look Up* is one example of how climate change can be imagined and communicated via a cultural text. The film not only represents culture (particularly when it comes to privileged nations and individuals) as one of the key obstacles to productively addressing climate change, but also shows that this very culture (responses to climate change and lack thereof) is bringing humanity and the more-than-human world to destruction and death.

When the PhD candidate in astronomy from the Michigan State University Kate Dibiasky discovers a comet, and her professor, Dr. Randall Mindy, confirms that the comet would hit the Earth in about six months, the two, together with the Head of the Planetary Defense Coordination Office at NASA, Dr. Teddy Oglethorpe, fly to Washington, D.C. to meet with the president and discuss what to do in order to prevent the catastrophe. The title of the film appears on screen exactly when Kate Dibiasky throws up into a trash bin, feeling overwhelmed because she would need to tell the president about the event that would most probably destroy the planet. Just as the title of the film suggests – *Don't Look Up* – Kate Dibiasky is looking down into the bin, which by now is filled with trash and vomit. Through this, the film anticipates the decision that many characters will make: to avoid confronting the issue. The trash can filled with garbage and vomit is a metaphor for denial, the war on science, and inaction, all of which are largely condemned in the film.

The scientists wait hours long until the president can meet with them, witnessing a serious of events, including a birthday party, that appear more important for the White House than the visit of the scientists. In the end, the president has time to meet them only next day. The contrast between the president and the people working for her on the one hand, and the scientists on the other, is visually established via the characters' clothes and accessories. The power of the politicians is emphasized via the way they are dressed. Dr. Mindy appears very stressed and nervous, trying to explain what catastrophic events will happen once the comet

---

27 Robert J. Brulle, "Denialism: Organized Opposition to Climate Change Action in the United States". In David Konisky (ed.), *Handbook of Environmental Policy.* Northampton: Edward Elgar Publishing, 2020, 338.

hits the Earth. The Chief of Staff, the president's son, pauses him, saying, "You're breathing weird. . . . It's making me uncomfortable." Dr. Mindy apologizes, saying that he is "just trying to articulate the science." To that, the man responds, with laughter, evidently undermining the importance of Dr. Mindy's message, "I know, but it's like so stressful, I like trying to like listen what..." Despite the warnings of the scientists that the comet is "a planet-killer" and it will cause "an extinction-level event," the administration decides to call it "a potentially significant event." When the scientists emphasize that the chances are "99.78 %" that the catastrophe will happen, i.e., this catastrophe is real and not just potential, the administration expresses great relief, thinking that as long as it is not a 100 %-sure case, the catastrophe can be avoided. The president even requests to "call it 70 % and . . . let's move on." The scientists are clearly perplexed with where this discussion leads to, to which the president reacts, with laughter: "You cannot go around saying to people that there is a 100 %-chance that they're gonna die. You know... it's just... nuts." She also suggests that other, allegedly more reliable, scientists should check this information, showcasing her distrust of the tenured professor of astronomy at Michigan State University, the doctoral candidate, and the head of the Planetary Defense Coordination Office at NASA, whose name she cannot even pronounce correctly. The concern here arises due to the fact that Kate Dibiasky and Dr. Mindy come from Michigan State which the Chief of Staff considers a bad university. He says to Dr. Oglethorpe, "Come on, bro", thus suggesting that everyone is implicitly informed about the fact that there are *better* scientific institutions out there with *better* scientists.[28]

The president ultimately interrupts the discussion, asking about the financial costs. Importantly, formulating her question as "What is this gonna cost me?", she suggests that she *alone* can solve this problem, and that *money* is the solution. The president remains calm, thereby trying to communicate to the scientists her viewpoint, namely that this is just another crisis that will take some time and perhaps demand some financial investment, but, in the end, just as multiple other problems that she had to deal with in the past, it will be solved. The president shares with the scientists: "You know how many the world-is-ending meetings that we've had over the years? Economic collapse, loose nukes, car exhaust killing the atmosphere, rogue AI..." The Chief of Staff continues: "droughts, famine, plague, alien invasion, population growth, hole in ozone." The comet is merely a new item on the list. Before the scientists leave, the Chief of Staff reminds them that this discussion is "super classified."[29] This suggests that interested individuals might keep certain

---

28 *Don't Look Up*, directed by Adam McKay. Los Angeles: Hyperobject Industries, 2021, Netflix.
29 *Don't Look Up*.

information secret if it directly affects them as well as that some of the politicians selected by the public to make decisions are incapable of performing their duties on a professional level, ultimately, risking lives of millions of individuals.

In this scene, the film comments on both the politics of denialism and the war on science, outlining them as forms of violence. First, the politicians refuse to trust reliable sources – the scientists. They were prejudiced from the very beginning, thinking that there simply could not be good scientists at Michigan State University. Dr. Mindy's and Kate Dibiasky's behavior only further confirms this to the politicians: Dr. Mindy appears too nervous and his explanations sound rather overwhelming and convoluted to the general public, whereas Kate Dibiasky, on the contrary, in the eyes of the administration, seems to be too aggressive. Second, politicians do not want to face the problem and, essentially, deny the fact that the planet can be destroyed and humanity will not survive this catastrophe. Despite all the facts and figures, the politicians insist that money can, yet again, solve everything. Denialism and the war on science, as depicted in this scene, are the forms of violence that lead to the destruction of the planet and the death of its human and nonhuman inhabitants.

Casualties: Climate Change Anxiety and the Problem of Responsibility

The violence that is depicted in *Don't Look Up* through denialism and the war on science inevitably leads to casualties. These casualties are the people whose professional opinion is neglected by the politicians, i.e., the scientists, but also all other humans, nonhumans, and the planet as a whole that die once the comet hits the Earth. Such representations directly reflect what Nixon identifies as "[c]asualties of slow violence" (both "human and environmental") that are "most likely not to be seen, not to be counted," and therefore are "light-weight, disposable casualties."[30] The politicians are fully responsible for the destruction that takes place at the end of the film. Certainly, there might have been no way to prevent the catastrophe, but the politicians neglected the problem at first, and later tried to address it using only the means through which they could potentially benefit. The plan is to extract resources from the comet, specifically minerals that could be used to advance technology. Those who are in power risk the future of the planet, humans and nonhumans, trying to make more profit. Here, the film suggests that humanity, and specifically the privileged, would rather die than live without the

---

30 Nixon, *Slow Violence*, 13.

comforts that they are used to. It is also crucial that in the film, the president of the United States, one of the most powerful countries in the world, makes this important decision for the whole world, thus silencing other nations.

Judith Butler's concept of "global responsibility" is particularly helpful to understand the problem outlined in *Don't Look Up*.[31] Butler claims that one way to perceive global inequality (via the example of war) is through the issue of life and living: "one way of posing the question of who 'we' are in these times of war is by asking whose lives are considered valuable, whose lives are mourned, and whose lives are considered ungrievable."[32] Similarly, Susan Sontag, examining war photographs, argues that these cultural artifacts are "a means of making 'real' (or 'more real') matters that the privileged and the merely safe might prefer to ignore."[33] In *Don't Look Up*, humanity, including Americans, is turned into causalities; yet this decision is made by the American president. This effectively mirrors today's global situation: scholars emphasize that people of color, indigenous people, and other minority groups experience environmental crises much stronger than the rest of the world yet do not have the financial means to address them.[34]

The violence generated by denialism and the war on science also produces or intensifies (eco-)anxiety. Harriet Dyer explains: "Eco-anxiety involves a complex mixture of feelings – including fear, frustration, grief, anxiety, helplessness, and hopelessness – but, in a sentence, it can be summarized as a feeling of dread about the future of the environment. It's a reaction to what's going on in the world and, for many people, it's actually a healthy response, as it shows that you are empathizing with the plight of the natural world."[35] (Eco-)anxiety is illustrated in the film through the behavior of Kate Dibiasky and Dr. Mindy. Both are overwhelmed by the knowledge they have; yet they become even more anxious when they find out that the government is not going to address the problem. Seymour writes, "knowledge is connected to affect, but in rather unexpected ways."[36] Drawing on research conducted by the environmental sociologist Kari Norgaard, Seymour explains that "the *more* one knows about climate change, the *less* one is likely to act"; "the real problem is not lack of knowledge or information but how the presence of knowledge or information can incite paralyzing emotions such as fear, helplessness, and guilt, which then make processing and acting on

31 Judith Butler, *Frames of War: When Is Life Grievable?* London: Verso, 2009, 36.
32 Butler, *Frames of War*, 38.
33 Susan Sontag, *Regarding the Pain of Others*. New York: Picador, 2003, 7.
34 Justyna Poray-Wybranowska, *Climate Change, Ecological Catastrophe, and the Contemporary Postcolonial Novel*. New York: Routledge, 2021, 60.
35 Harriet Dyer, *Eco-Anxiety (and What to Do about It)*. London: Summersdale Publishers, 2020, 4.
36 Seymour, *Bad Environmentalism*, 46.

that knowledge or information difficult or even impossible."[37] It is important to clarify here, however, that in the film, characters like the president deny scientific facts and thus exclude them from what they consider knowledge. Their knowledge is that even if the catastrophe takes place, they will be able to escape it. This helps them stay calm until the very end. They do not necessarily reject the catastrophe, but rather, they do not care what that catastrophe will do to those who have no money, power, or influential friends.

Ailise Bulfin contends that the multiple images of catastrophe generated by cultural texts, including non-environmental catastrophes, emerge because of "fears of human-induced ecological disaster, and particularly climate change."[38] In other words, the end of the world that these images visualize is "indicative of the reality of the new human condition of disruption to the very environment which sustains our existence."[39] Fear is generated by the reality that is informed by multiple crises, including, most prominently today, the climate crisis and health crisis. But fear is also instigated and used as a weapon to control and manipulate the public by those who deny climate change (and other problems like, for example, the coronavirus pandemic). When it comes specifically to climate change, it is very problematic to think of fear as an emotion that can trigger action and productive change. Rachel A. Howell claims that "[c]ampaigns which appeal to fear as a motivator are problematic because fear can trigger denial, apathy, repression, anger, and counterproductive defensive behaviors."[40] In turn, Philip Hammond and Hugh Ortega Breton argue that "when dealing with future-located risks such as major anthropogenic climate change, the use of fear of catastrophe appears to be ineffective because the severity of predicted dangers does not correlate with everyday knowledge and experience of present circumstances."[41] Hence, using fear to promote pro-environmental action can not only prevent one from changing their anti-environmental behavior or help minimize the effects of the ongoing climate crisis, but also distort the very understanding of the crisis in people's minds. Fear is a natural psychological mechanism; it is a reaction that, in the context of climate change, signifies one's heed and care. However, instigating fear or viewing fear as the only motivator for action is wrong and counterproductive. Fear can sabotage necessary action, which is effectively illustrated through the two sci-

37 Seymour, *Bad Environmentalism*, 45. [Italics in original.]
38 Bulfin, "Popular Culture and the 'New Human Condition'", 142.
39 Bulfin, "Popular Culture and the 'New Human Condition'", 142.
40 Rachel A. Howell, "Lights, Camera . . . Action? Altered Attitudes and Behaviour in Response to the Climate Change Film *The Age of Stupid*". *Global Environmental Change* 21 (2011): 178.
41 Philip Hammond and Hugh Ortega Breton, "Bridging the Political Deficit: Loss, Morality, and Agency in Films Addressing Climate Change". *Communication, Culture & Critique* 7 (2014): 303.

entists, Dr. Mindy and Kate Dibiasky, who fail to communicate the serious nature of the problem to the public, not least because of the fear that wears them out. It is important to note, however, that the film does not suggest that the scientists' fear is the reason why nothing has been done to prevent the catastrophe or at least minimize its effects. The indifference of the politicians and general public played a much more significant role in allowing this catastrophe to happen. Yet fear is depicted as one emotion that can be detrimental, for it causes chaos, hatred, apathy, and other emotions that can prevent individual and collective action.

Unlike the president, Kate Dibiasky and Dr. Mindy cannot remain calm. They have the knowledge and understand how serious the situation is, and their first reaction is anxiety. Dr. Mindy tries to influence other people by sharing the knowledge that he has, yet as soon as he realizes that no one listens to him, he gives up. Once a good husband and a family man, he starts cheating on his wife with a news reporter. Kate Dibiasky's reaction is different. She is depicted as being nervous; it is difficult for her to control her emotions. In one of the shows where she and Dr. Mindy tell about the comet, Kate Dibiasky explodes: "Maybe the destruction of the entire planet isn't supposed to be fun. Maybe it's supposed to be terrifying, and unsettling, and you should stay up all night, every night, crying, when we're all 100% for sure gonna fucking die!" Crying, she runs out of the studio. The hosts turn the situation into a joke, emphasizing that Kate Dibiasky should have taken some Xanax, which would have helped her stay calm. This makes her appear as a mentally unstable woman, discrediting her both as a woman and as a scientist. The comet is named after the doctoral candidate (because Kate Dibiasky discovered it), and the plan to explode the comet can be interpreted as the desire to get rid of the woman scientist. Once the information about the catastrophe becomes public, the president even asks people, "don't blame her," thus suggesting that some individuals might think that the catastrophe is Kate Dibiasky's fault.[42]

Kate Dibiasky experiences (eco-)anxiety and cannot cope with her feelings for a long time. She feels desperate and betrayed by the government, the state, as well as her family and friends who do not want to have anything to do with her, believing that she has gone mad. In *A Field Guide to Climate Anxiety*, Sarah Jaquette Ray states, "those of us who feel the most passionate about social change are also the most likely to fail to be considerate of ourselves."[43] This is exactly what happens to Kate Dibiasky, who hopes to change humanity's actions and behaviors within six months. Yet Kate Dibiasky witnesses that social media is more interested in the

---

42 *Don't Look Up.*
43 Sarah Jaquette Ray, *A Field Guide to Climate Anxiety: How to Keep your Cool on a Warming Planet.* Oakland: University of California Press, 2020, 128.

love story of two celebrities who have temporarily separated than in the global cat-astrophe that will destroy the planet and kill every human and nonhuman living on it. The fixation on the world-as-we-know-it and the insignificant issues, such as the private lives of celebrities, that capitalism has introduced as essential compo-nents of good western life is ridiculed, and humanity is portrayed as degraded and no longer capable of distinguishing between significant problems and insignificant ones and prioritizing the former. People are consumers that thoughtlessly and au-tomatically do things that are characterized by producers as worth having and spending time on. To further support this claim, the film includes an iconic scene of Ariane Grande's character and her boyfriend performing while humanity is preparing to die. The baseball caps reading "Don't Look Up" that those who join the event wear are reminiscent of Trump's "Make America Great Again" caps, which can be interpreted as a reference to denialism promoted by populists and what it can lead to.

Crucially, the film speaks about (eco-)anxiety from the perspective of predom-inantly white characters who live in the United States. *Don't Look Up* is another example of apocalyptic cinema that generates what E. Ann Kaplan defines as "pre-trauma," which in part is characterized by "severe anxiety about the future in Eu-rocentric cultures."[44] Dealing with the unavoidable catastrophe, which can be read as a metaphor for the current environmental crisis, the characters fail to address this problem because it appears too complex and demands too big a change. As such, one can view this catastophe as a "[h]yperobject," the term that Timothy Mor-ton uses to speak about "things that are massively distributed in time and space relative to humans."[45] Everything that the characters are concerned with is that the world-as-we-know-it is coming to an end. There is hardly any worry about the most vulnerable ones or the destruction of the planet as such. Examining the current climate crisis and climate anxiety, Blanche Verlie urges us to "learn to live with climate change":

> Learning to live with climate change begins from an understanding that *climate is living-with*... Climate as living-with focuses on the intimate ways we are entangled with the non-human world, and how the patterns of these relationships generate the conditions in which we live. It therefore attunes to how the planetary and epochal phenomenon of climate change is metabolically, emotionally and politically enmeshed within our everyday, mundane, inter/personal lives and compels respect, reciprocity and responsibility for this expansive re-lationality. This understanding better enables us to cultivate strategies that can adequately in-

---

44 Kaplan, *Climate Trauma*, 1.
45 Timothy Morton, *Hyperobjects: Philosophy and Ecology after the End of the World*. Minneapolis: University of Minnesota Press, 2013, 1.

spire and support people to engage with and respond to climate change, and to do this in ways that contribute to multispecies climate justice.[46]

Most importantly, Verlie emphasizes: "I want an approach that does not help privileged people feel okay or help them accept climate change, but one which enables us to bear the burden of complicity in ways that hold us accountable and that generate radical change."[47] This feeling and realization of "bearing the burden" and productive actions to address the problem through change are both missing in the film. As *Don't Look Up* illustrates, humans either do not feel guilty and responsible for anything, or see the problem but do not do anything, or insufficiently contribute to help solve the crisis.

In the middle of the credits, the film depicts a scene that happens 22,740 years later. The privileged, including the president, her administration, the billionaire who suggested extracting minerals from the comet, and a few other lucky ones, survived the catastrophe on a ship. They wake up on a planet that looks like it was during the Mesozoic era. The characters emerge naked from the ship, like new Adams and Eves that have arrived to inhabit the new place. The president is soon eaten by one of the creatures living on the planet; allegedly, the same might happen to the rest of the group. The film ends with the image of the president's son, who has survived the catastrophe. Amidst destruction, he is recording a video for social media and asking his potential viewers to like the video and subscribe to his channel. Ending with this scene, the film illustrates how corrupt and essentially wrong humanity's priorities have become, and how disinterested humanity is in what is happening to the planet right now as well as what the future will look like. Today we are, like the president's son, surrounded by destruction – the ongoing environmental crisis – yet many of us refuse to see it and continue to hope that we and the planet as a whole will survive no matter what.

Conclusion

Violence has gained a new meaning in today's world. Denying climate change, among other life-threatening issues, is a form of violence. Doubting science and sabotaging the dialogue between the scientific community and the public is a form of violence. The quotations from Brunner's *The Sheep Look Up* that open this essay, juxtaposed with the scenario sketched out in *Don't Look Up*, become par-

---

46 Verlie, *Learning to Live with Climate Change*, 5. [Italics in original.]
47 Verlie, *Learning to Live with Climate Change*, 8.

ticularly important today, when humanity has to address the environmental crisis. Why does humanity continue to neglect environmental problems? Is it because there is not enough environmental knowledge and environmental education in general? Or is it because humanity has collectively entered the phase of eco-anxiety, realizing that it is hard to productively deal with certain issues? Bronislaw Szerszynski argues, "The persistence of unsustainability is due not simply to the ignorance or duplicity of individuals, or even to the mere logic of the capitalist system, but also to a crisis in political meaning in which we are all implicated. The solution… is not to be found in a simple restoration of political language's reference to a reality outside language, as if language is a flapping sail that can simply be re-secured to its mast."[48] Can humanity adopt pro-environmental thinking, politics, and change culture to such an extent that would help humans solve the ongoing environmental crisis? Grove argues that to think that climate change would unite people by "provid[ing] a universal ground for the cosmopolitan solidarity as-yet unachieved by other means is dangerously naïve."[49] Humanity continues to be disintegrated; it fails to address global crises (including the ongoing pandemic) together, outlining a set of rules that would apply to the whole world rather than to specific nations. Such an approach is dangerous and unproductive and must be changed. There are multiple ways in which violence manifests itself today, and being fixated on the world-as-we-know-it and developing policies and different strategies aimed at the preservation of this concept is wrong and counterproductive. The health of the environment is of crucial importance and should be placed at the center of attention.

---

48 Bronislaw Szerszynski qtd. in Seymour, *Bad Environmentalism*, 43–44. [Italics in original.]
49 Grove, *Savage Ecology*, 36.

Stefania Paladini

Chapter V
Unsustainable Wars? The Use of Weapons in Lower Earth Orbit

The recent Russian ASAT (anti-satellite) missile test of November 2021, which alleg-edly generated more than 1,500 trackable orbital debris and forced the ISS astro-nauts to seek refuge, is only the last of a series of wargame tests held in LEO (the Earth's Lower Orbit). And while ASAT tests are not the only ones responsible for the alarming growth of debris in Earth's orbit, each instance exponentially worsens what is now a clear menace to space activities even before it constitutes a security issue in military terms. The growing number of space actors only makes this state of things more urgent to address, given the willingness of countries to test their own ASAT missiles (as the not-so-recent examples of China and India prove). Part of the problem is, of course, the legal framework for the use of weap-ons in space, which explicitly forbids WMDs but not ASAT. Taking stock of the pre-sent challenges in regulatory terms but also the growing competition in outer space, this article explores the risk of unregulated use of unconventional weapons in space and the necessity of making sure space remains the 'province of human-ity' to be used for the common benefit, as it has always intended to be.

A Contested and Challenged Environment

In one of its recent (and increasingly frequent) policy statements about space, the EU declared that "space is becoming a more contested and challenged environ-ment." New competitors – both public and private – are emerging around the world, partly spurred by the reduced cost of developing and launching satellites. Growing threats are also emerging in space: from space debris to cyber threats or the impact of space weather.[1] It is impossible to disagree: the growing number of space actors, both private and public, and the sheer number of endeavours re-lated to some titles as the 'last frontier', only make those challenges and threats more urgent to address. After all, from being a niche industry with limited mon-

---

1 Communication from the Commission to the European Parliament, the Council, the European Economic and Social Committee and the Committee of the Regions. Space Strategy for Europe (COM 2016/705 Final), Brussels. 2016.

https://doi.org/10.1515/9783111081687-006

etary value (albeit impressive potential), the space sector has now grown to a considerable size.

The space economy has continued to grow rapidly during the pandemic, and the Bank of America estimated in 2020 that the growth would more than triple in size in the next decade, to become a $1.4 trillion market. 2021 was the biggest year on record for private investment in the sector (US $4.5 billion only in the second quarter of the year, according to Morgan Stanley), with more space launches (145, 133 of which were successful) than ever before in history.[2]

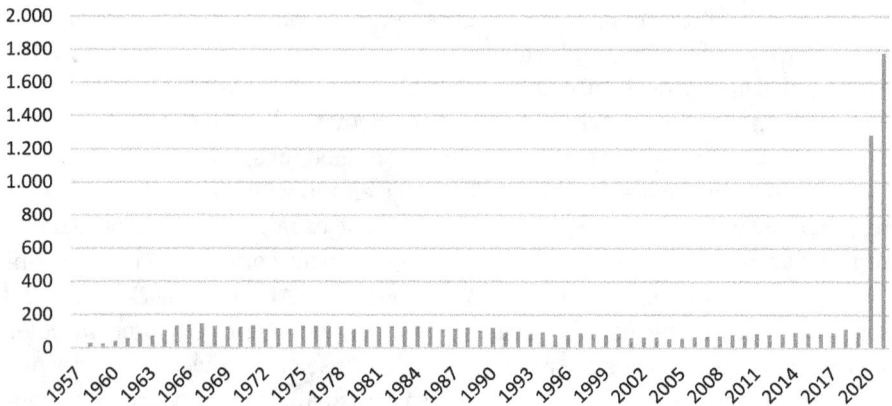

Figure 1: Number of satellites launched by year 1957–2020.
Source: Author's elaboration on SIA, Statista, 2022.

This should not come as a surprise: more than 80 countries have so far launched their own satellites, and launching facilities are rapidly growing, with more spaceports being built now than at any time before. According to the Satellite Industry Association (SIA), there were 3,371 satellites orbiting Earth by the end of 2020, an increase of 37% compared to 2019.[3] The space industry is booming, and not just in the West. In China, there were about 60+ privately funded start-ups in 2021. said that the impressive growth of commercial space operations should not make us forget how the defence and military components still play an important role.

---

2 Statistics for 2020 and 2021 are high-skewed due to the launch of satellite constellations, such as OneWeb and Starlink.
3 SIA. State of the Satellite Industry Report 2021 at https://sia.org/news-resources/state-of-the-satellite-industry-report. Incidentally, and just for information, UNOOSA statistics differ, and they are considerably higher, because they account also for non-active satellites. UNOOSA, 2022.

Space Security or Security in Space?

From the onset of the space age in 1957, when the Soviet Union launched Sputnik 1 into orbit, space and security have been intertwined in terms of aims, means, and modalities. In some countries, they still are, and so is the space sector, where commercial activities are often indistinguishable from military endeavors and funded in the same way. Which makes increasingly difficulty separating commercial, space-enabling capabilities from launching capabilities, which rely on the same technology used for ballistic missiles.

According to some estimates, the United States and the Soviet Union/Russia have launched in between 1960 and 2015 about 2000 military satellites. Even now, about 20–30 % of some 1000 satellites that are currently in orbit are directly related to such security functions and are usually operated by military institutions'.[4] As a result, even now national security is an inherent feature of the space industry[5] and, as it stands, there is a growing corpus of academic literature that deals with the various aspects of this complex relationship, taking also into account how the increasingly civilian presence in space has influenced the rules of the game.

The ongoing debate of security is still heavily influenced by a realist ideology[6], which has largely mirrored the existing geopolitical canon of sea power doctrine by Alfred Thayer Mahan[7] in 1899, adopting the same six fundamental elements (geographical position, physical conformation, extent of territory, size of population, character of the people, and character of government) identified for high seas to analyze space politics[8]. Dolman has also suggested that "geopolitical theory developed for the Earth and its geographical setting can be transferred to outer

---

4 Karl Schrogl, Peter L. Hays, Jana Robinson, Denis Moura and Christina Giannopapa (eds.), *Handbook of Space Security*. New York: Springer Media, 2015, 14.
5 Ram S. Jakhu and Joseph N. Pelton. *Global Space Governance: An Inter-national Study*. Cham: Springer International, 2017, 4.
6 Robert Pfalzgraff, *International Relations Theory and Space Power*. London: Institute for National Strategic Studies, 2015.
7 Alfred Thayan Mahan, *The Influence of Sea Power Over History*, 1890. Available online at: https://archive.org/details/seanpowerinf00maha
8 Francis P. Sempa, "The Geopolitical Vision of Alfred Thayer Mahan". *The Diplomat*. 30 December 2014, accessed 2018. Available online at: https://thediplomat.com/2014/12/the-geopolitical-vision-of-alfred-thayer-mahan/. Michael Leissle, Thomas Hoerber, and Gabriel Siglas (eds.), *Power, Politics and the Forma tion of International Law: a Historical Comparison, Theorising Space*. London: Lexington, 2017.

space" with the "strategic application of new and emerging technologies within a framework of geographic, topographic, and positional knowledge"[9].

The same can be said about the eventuality of a war in outer space. Contrary to popular belief and Star Wars scenarios, and as Gen. John Hyten of the US Strategic Command rightly noted, there is "no such thing as a war in space, there is just war, and it's with an adversary and if it extends into space, we have to figure out how to fight it."[10] It is not by mistake that what has been labelled "the first space war" – i.e., the 1991 First Gulf War – did not happen in space at all.[11] but made extensive use of GPS (Global Positioning System) and other satellite capabilities to direct military operations on Earth. And what was true in the 1990s is even truer today, when the military capabilities of major powers – the US included – are increasingly dependent on space-based facilities, making the likelihood of asymmetric attacks on them an important issue.

This is not the place to elaborate on the concept of space security and how it has expanded from what was "a two-dimensional model of military and environmental dimensions of space security"[12] to a more complex and multifaceted definition that does not only include only non-state actors, e.g., private operators, but also non-traditional kind of threats originating in space, both man-made (debris) and not, such as space weather, space contamination and asteroids among them[13]. Or to discuss how, rhetoric aside, there is no such thing as a "Pearl Harbour" in space that could instantly cripple any major power's satellite network.[14] What matters here is that nation states remain the dominant actors in outer space for the time being, and thus the use of space for military purposes, alone or in conjunction with more civilian, commercial-oriented uses. is, of course, at the root of the issues analyzed in this article: space weapons and their impact on the sustainability of space.

9 Everett Dolman, *Astropolitik: Classic Geopolitics in the Space Age.* London: Frank Cass, 2002, 15.
10 Arthur D. Villasanta, "US will fight and win a space war – and is preparing for it". chinatopix.com, 27 Feb 2017, accessed 18 April 2022. Available online at: http://www.chinatopix.com/articles/111937/20170227/will-fight-win-space-war-preparing.htm
11 Peter Anson and Dennis Cummings, "The First Space War: the contribution of Satellites to the Gulf War". *RUSI Journal* 136 (1991): 45–53.
12 Martin Sheehan, "Defining Space Security". In Kai-Uwe Schrogl et al. (eds.), *Handbook of Space Security.* New York: Springer-Verlag, 2015.
13 ESA. Latest Report on Space Junk. 2018.
14 The US Government. Report of the Commission to Assess United States National Security Space Management and Organization, 11 January 2001. In Geoffrey Forden, "Viewpoint: China and Space War, Astropolitics", *The International Journal of Space Politics & Policy* 6:2 (2008): 138–153.

The Debate About ASATs: Issues and Legitimacy

Among the wider debate about space security, there is one particularly complicated in terms of its implications for the commercial space sector: space weapons. The existence of space weapons itself is not a novelty. Earth's orbits have been exploited for military purposes since the beginning of the space age. And in itself, the development of anti-satellite weapons has, maybe surprisingly, not led (so far) to the weaponisation of space. As Hostbeck correctly states, "the path of conflict in space and through space starts and ends on the ground." However, as long as ground conflicts make use of space-bound technology, there's always the risk that the conflict could evolve in character and space-specific weapons will be developed and deployed.[15]

To understand the complexity of space weapons means examining how the existing legal framework deals with them. This is because regulations are actually part of the problem that weapons in space represent, not of the solution. Part of the complexity is due to the fact that the debate about what constitutes a "space weapon" is as old as the debate about space security itself, with a lot of confusion and contrasting views about it.[16] No official definition exists to date, and the equation nuclear weapons=space weapons is flawed for more than a reason.[17]

There are more than a few spacecraft and probes of various kinds that can be defined as "dual use," meaning that they can be used both in commercial and defensive operations. There are a few examples worth analyzing for their far-reaching implications, such as Aolong-1, the probe China launched in 2016 with the specific – and laudable-task of cleaning Earth's orbits from the growing issue of debris (see further for this). However, there were a few claims that the spacecraft had other functions as well, equipped as it was with satellite targeting capabilities.[18] Another case is that of the so-called "inspection satellites," equipped with tracking and disabling capabilities.[19]

15 Lars Hostbeck, "Space Weapons' Concepts and Their International Security Implications". In Karl Schrogl, Peter L. Hays, Jana Robinson, Denis Moura and Christina Giannopapa (eds.), *Handbook of Space Security*. New York: Springer Media, 2015,970.
16 Stefania Paladini, *The New Frontiers of Space*. London: Macmillan, 2019.
17 Karl D. Hebert, "Regulation of Space Weapons: Ensuring Stability and Continued Use of Outer Space". *Astropolitics* 12 (1) (2014): 1–26.
18 Harsh Vasani, "How China is weaponizing outer space". *The Diplomat*, 19 January 2017, accessed3 January 2018. Available online at: https://thediplomat.com/2017/01/how-china-is-weaponizing-outer-space/
19 Linda Dawson, *War in Space*. London: Springer, 2018.

Finally, there are the so-called ASAT (anti-satellite weapons), probably the most talked about spacecraft in a security concept. Originally developed by the then-USSR in the 1960s and made fully operational in the 1990s,[20] they are probes sent to the same orbit as the target satellite armed with some kind of offensive weapon (kinetic, mechanical, laser, etc) that can either destroy or just disable the target. In the event of a conflict, they might prove a real threat because, once ASATs have been launched against an orbiting target, it would be difficult (for the targeted satellite) to avoid the collision.

A different but related issue is what involves the legal aspects of the matter. For what strictly concerns the internationally applicable space law, the outlook in 2021 is still the same as in 1967, even though the landscape has dramatically evolved since those days. And if ASAT is what most qualifies as a proper space weapon, rather than just a "weapon used in space," then the way international space law and treaties are written will change[21] which will lead to a systematic but more complex framework.

In short, international treaties do allow ASAT deployment. "Existing laws do not provide adequate restrictions on space weapons. While they do restrict weapons of mass destruction (WMD), everything else is allowed'[22], ASAT included, as all the treaties signed so far do allow the use of force in a defensive function and therefore the use of weapon in space, except what is explicitly forbidden (i.e., WMDs). As a matter of fact, there is a big difference between putting arms into space and "weaponizing space", and this is where the disagreement between the USA on one hand and Russia and China on the other resides. It also shows the limits of the existing international treaty framework, which was created in the 1970s but is proving no longer apt to deal with the present situation.

For what concerns specifically ASAT, there are only a handful of countries in possession of them to date – namely, the USA, China, Russia, and India. All of them have used ASAT as direct-ascent missiles to strike a satellite by using a tra-

---

20 Anatoly Zak, "The hidden history of the soviet satellite-killer". *Popular Mechanics*, 1 November 2013, accessed 20 December, 2017. Available online at: http://www.popularmechanics.com/space/satellites/a9620/the-hidden-history-of-the-soviet-satellite-killer-16108970/

21 For the sake of clarity, what is generally referred to as "space law" is actually made of five international treaties, whose application the UNOOSA (the UN agency Office for Outer Space Affairs) oversees. The most important and the cornerstone of the entire framework is The Outer Space Treaty" (OST) 1967. OST states the cardinal principles that regulate the use of space ('space as a province of mankind', 'free for everybody use'; Art. 1 and Art 2) and paving the way of demilitarization of outer space promoting its peaceful uses for everybody (art 3).

22 Hebert, "Regulation of Space Weapons", 4.

jectory that intersects its target without placing the interceptor in orbit.[23] All of them have used one of their own assets as a target for the test, to avoid triggering a direct military confrontation, which, however, did not spare them from creating international issues nonetheless (see further).

Controversies apart, what made things complicated is that it was in the countries' right to carry out the tests: there are no regulations that specifically restrict or even address ASAT use, although it has often been attempted at international level. Among them, the most substantial was the proposal of an arms control agreement for outer space sponsored by China and Russia in 2008, the so-called Draft Treaty on the Prevention of Placement of Weapons in Outer Space and the Threat or Use of Force Against Space Objects (PPWT), revised in 2014. The treaty called for an outright ban on the "placement of weapons in outer space as one of the most important instruments of strengthening global stability and equal and indivisible security for all."[24] However, this initiative was stopped by the US refusal to back the treaty on the valid ground that the treaty was not dealing with the ASAT weapons and, therefore, not addressing the most significant threat of weapons in space. Since then, no progress has been made, while the ASAT tests in orbit have continued.

ASAT Tests and Their Geopolitical Implications

The continuing testing of ASAT proved the US right in its refusal to commit to any treaty which does not regulate them in the first place. On November 14, 2021, Russia tested an ASAT (anti-satellite) missile in LEO (the Earth's Lower Orbit) targeting its own Cosmos-1408 satellite – the latest of the countries with ASAT capabilities to carry out such a test, whose consequences in terms of the space environment will be discussed in the following section. The Russian test followed a decade of highly controversial ASAT tests, which were started by China. In January 2007, China conducted an anti-satellite (ASAT) weapons test and destroyed its own satellite Fen-

---

23 Almudena Ortega, "Placement of Weapons in Outer Space: The Dichotomy Between Word and Deed". *Lawfare*, 28 January 2021. Available online at: https://www.lawfareblog.com/placement-weapons-outer-space-dichotomy-between-word-and-deed
24 VOA News. "U.S to Confront Russia, China in Militarization of Outer Space", 4 October 2018, accessed on 3 January 2021. Available at: https://www.voanews.com/a/us-to-confront-russia-china-on-militarization-of-outer-space/4599793.html

gyun-1C, creating an outcry in the space community[25], for the debris it created (see further).

The test served to consolidate China's position as the third big military player in outer space, together with Russia and the USA. The US military analysts from NORAD reconstructed the collision dynamics, proving that the Chinese ASAT performed a sophisticated, high-precision maneuver, roughly equivalent to hitting a bullet with another bullet, and was, therefore, very advanced in terms of technology, superior to, for instance, their Soviet-era counterparts.[26]

Thus, in addition to the debris, the Chinese test provoked what can be considered a good example of the security dilemma (progressive escalation of threats due to tit-for-tat)[27], fuelled by the obvious but popular misconception of space as an 'offence-dominant environment'.[28] It didn't help that, after ASAT testing, China carried out different but still military-relevant tests (missile defence) in 2010 and again in 2013. 2013 was held close to geosynchronous orbit, where the majority of ISR (Intelligence, Surveillance, and Reconnaissance) satellites are located. Yet another test was performed in 2015, this time an exo-atmospheric probe with on-board capabilities to destroy satellites.[29]

Among the countries that felt more directly threatened, there was, obviously, India, in yet another example that geopolitical equilibria on Earth do matter in space, too. About one decade after China's ASAT testing, and in direct response to that, India tested its own first ASAT, too, in April 2019, "Mission Shakti" and re-established an equilibrium between the two Asian powers in outer space. The April 2019 test was also important from another perspective because it gives strength to an ongoing trend already existing in India, and namely, moving from a space sector traditionally civilian based and ruled[30] to one where, on the Russian and China's tradition, far from influenced and intertwined with the defence com-

---

25 To be true, the 2007 ASAT test was not the only one China carried out. There were failed attempts before that, and other following it. However, the January 2007 one is considered a watershed, both in terms of a landmark of space capabilities and in adverse consequences it created.
26 Forden, "Viewpoint: China and Space War, Astropolitics".
27 John H. Herz, "Idealist Internationalism and the Security Dilemma". *World Politics* 2, no. 2 (1950): 157.
28 Brad Townsend, "Strategic Choice and the Orbital Security Dilemma". *Strategic Studies Quarterly*, 14, no. 1 (2020): 64–90.
29 Tania Branigan, "China successfully tests missile interceptor". *The Guardian*, 12 January 2010. Available online at: https://www.theguardian.com/world/2010/jan/12/china-tests-missile-interceptor
30 ISRO (the Indian national space agency) has traditionally treated the military as one of its (many) stakeholders and allowing no direct influence over the agency's governance or operations. Paladini, *The New Frontiers of Space*.

munity[31]. But even more important than the geopolitical implications were the systemic costs of the test.

The Environmental Costs of ASAT Testing

Apart from triggering the well-known security dilemma dynamics, ASAT tests come with a high price in environmental terms. (Space) debris is a wide-encompassing term which designates all the celestial objects, small and big, such as fragments of satellites, rocket parts, remains of explosions or collisions, which orbit around the Earth at different altitudes. Depending on their size, location, and composition, they present a certain kind of threat. The risk posed by debris is of increasing concern, and a few international space agencies are tracking their number; NASA, JAXA, and ESA are among them.

Now, while ASATs are not the only responsible of debris formation – at least 200 instances have been recorded of satellite explosions due to malfunctioning, collisions (e. g. Iridium 33 case, discussed in section 7), or aborted launches.[32] Each instance of an ASAT test exponentially worsens what is now a clear menace to civilian space activities even before constituting a security issue in military terms. There is plenty of evidence in this sense. According to the ESA maintained database (SDUP), while only 11% of fragmentation in space is due to intentional events (like ASAT), their contribution to debris production is relatively higher than other causes.

Table 1: Top 10 fragmentation events by number of fragments (1957–2022).[33]

Rank	Name	Int. designator	Breakup date	Fragments
1	**Fengyun 1C**	**1999–025 A**	**2007–01–11**	**2809**
2	**(Cosmos 1408)**	**1982–092 A**	**2021–11–15**	**1500**
3	*Cosmos-2251*	*1993–036 A*	*2009–02–10*	*1070*

---

31 James C. Moltz, *Asia's Space Race: National Motivations, Regional Rivalries, and International Risk.* New York: Columbia University Press, 2012.
32 Anne Lemaître. "Space Debris: From LEO to GEO." in *Satellite Dynamics and Space Missions,* eds. by Giulio Baù et al. (New York: Springer 2019), 115–157.
33 The original ESA table covered debris production only up to end of 2918 and did not include either Indian or Russian ASAT. As if they were included, they would be among the top ten list, they have been added by the author within brackets.

Table 1: Top 10 fragmentation events by number of fragments (1957–2022). *(Continued)*

Rank	Name	Int. designator	Breakup date	Fragments
4	NOAA 16	2000–055 A	2015–11–25	457
5	Centaur-5 SEC	2014–055B	2018–08–30	455
6	Cosmos-1275	1981–053 A	1981–07–24	419
7	**(Microsat-R)**	**2019–006 A**	**2019–01–24**	**400**
8	*Iridium 33*	*1997–051C*	*2009–02–10*	*326*
9	Agena D	1970–025C	1970–10–17	232
10	DMSP Block 5D-2 F13	1995–015 A	2015–02–03	218

Source: Author's elaborations on ESA dataset, 2021

The second most impactful event in terms of debris production was also the latest ASAT test recorded to date, i.e., the November 2021 Russian test, which generated more than 1,500 trackable orbital debris and forced the ISS astronauts to seek refuge from the 'debris cloud'.[34]

Similar wargame exercises and tests held in LEO (the Earth's Lower Orbit) have been routinely carried out by spacefaring countries, adding to the other causes (accidental collisions, launching failures, and others). Not all of them, however, had the same impact in terms of consequences, as Chart 2 illustrates. According to ESA data, the Chinese ASAT test was, to date, the worst debris-producing event to date since the start of the space age.[35] The target was the polar-orbiting weather satellite Fengyun-1C "Wind Cloud" (FY-1C), launched into sun-synchronous orbit on May 10, 1999. The satellite was initially designed to stay in orbit for just two years. However, its working time was extended until 2005.

The ASAT, carried over in 2007, test was performed at an altitude of 865 km and produced a debris field of some 3,000 objects[36] that will linger in space for decades and constitute a hazard for all the orbiting satellites and the International Space Station. Furthermore, there is mounting evidence that debris from the test may

---

34 It was not the first time for the ISS to have perform escape manoeuvres due to debris. A few examples have been reported in the press (BBC, 2012). This is expected to grow in parallel with the increasing debris presence.

35 Conventionally set on 4 October 1951, when the Soviet Sputnik 1 was sent into orbit.

36 Number estimates vary a great deal, depending on the size of debris taken into account, from about 2800s (ESA data) to 3378 (NASA figures). But the exceptionality of the damage is not under discussion.

well fall to Earth and cause additional damage, implying that the dangers are not limited to orbit. Compared to the Chinese and Russian ASAT tests, India's had comparatively less impact, which is the reason why India claimed it had acted "responsibly" when performing Sakthi. While many have questioned it, such a claim has some validity. The test was performed at a low altitude, below 300 km, in order to avoid creating debris at the altitudes of operational satellites in low earth orbit (many of which orbit at altitudes between 400 and 1200 km). As the target was a small satellite (Microsat-R), only about 400 debris objects resulted from the collision, 270 of which are still being monitored by the Indian Space Agency. Still, as shown in Table 1, the test is still at #7 in the ranking presented in Chart 1, which is rather high for an event presented as "low impact."

The science behind debris production is well known, and no-one performing ASAT tests ignores it. Debris is due to the so-called "Kessler effect", a chain reaction of collisions that makes every accident raise exponentially the number of fragments in Earth's orbit, and which risks making some of those orbits no longer viable for human activities of any kind.[37] With about 20,000 objects larger than 10 cm, which can significantly harm a satellite, catalogued and tracked by NASA, and the growing likelihood of an increase in their number, there is all the reason to start thinking of a sustainable solution.

Space Environment as Non-Traditional Security. Issues to Address

ASAT tests are, of course, not the only responsible for the alarming growth of debris in Earth's orbit, as Table 1 showed. What is certain is that the growing number of space actors only makes this state of things more urgent to address, given the willingness of countries to test their own ASAT missiles.

Debris is an environmental security, which is one of first to be identified as a non-traditional security issue[38] and as a geopolitics component[39]. Scarcity is re-

---

37 Donald J. Kessler and Burton G. Cour-Palais, "Collision Frequency of Artificial Satellites: The Creation of a Debris Belt". *Journal of Geophysical Research* 83(6) (1978): 2637–2646.
38 Barry Buzan and Ole Waever, *Regions and Powers: The Structure of International Security.* Cambridge. UK: Cambridge University Press, 2003. Daniel H. Deudney and Richard A. Matthew, *Contested Grounds: Security and Conflict in the New Environmental Politics.* Albany: Albany State University Press, 1999.
39 Lorraine Elliot, "Regional Environmental Security: Pursuing a Non-Traditional Approach". In Andrew T. H. Tan and J. D. Kenneth Boutin (eds.), *Non-Traditional Security Issues.* Singapore: Institute of Defence and Strategic Studies, 2002.

garded as the fundamental problem[40], together with the related aspect of the access to resources is that link to conflict in terms of crowded orbits for new space actors, the way scarcity on Earth has historically created resource wars. The risk is substantial and not to be underestimated. Space might well be infinite, but usable Earth orbits to lodge satellites is not. The three most commonly used Earth orbits are: Lower Earth Orbit (LEO), the closest to the planet, Medium Earth Orbit (MEO), and Geosynchronous Orbit (GEO), are the most commonly used ones, and their extensions are limited. Moreover, these limited areas have been thinning for decades, and they are increasingly at risk of running out. ITU has started warning states about the overpopulation of GEO since the beginning of the 1970s.[41]

Malfunctions due to the overcrowding of Earth's orbits have been happening for decades. In 1996, the United Nations reported that severe crowding in the geostationary orbital slots over Asia "led to the jamming of a communication satellite by PT Pasifik Satellite Nusantara (PSN) of Jakarta, Indonesia, in defence of an orbital position claimed by Indonesia." This incident focused global attention on the worsening problem of orbital crowding and caused the matter to be brought before the 1997 World Radio Communication Conference (WRC) of the 187 member-nation ITU in Geneva.[42] In this scarcity-threatened landscape, ASAT-related debris complicates it even more. An example of what can happen has been exemplified in the now infamous Iridium 33 collision case.

Legal Implications of Debris: The Case of Iridium 33-Cosmos 2251 Collision

Another calamitous event, ranked as # 3 and # 8 in Table 1 in terms of debris production, was a collision, which is worth studying for the far-reaching legal implications of the case. On February 10, 2009, there was the first accidental collision between two satellites, the derelict-but-still-functioning Russian satellite Cosmos 2251

---

40 Stefania Paladini, "Resource Controversies Between Conflict and Cooperation: Agent-Based Models for Non-Traditional Security". In Martin Neumann and Davide Secchi (eds.), *Agent-Based Simulation of Organizational Behavior.* Cham: Springer, 2015. Michael Klare, "The New Geography of Conflict". *Foreign Affairs* May 2001.
41 ITU. Optional Protocol on the Compulsory Settlement of Disputes Relating to the Constitution of the International Telecommunication Union, to the Convention of the International Telecommunication Union and to the Administrative Regulations, 1992.
42 United Nations, Highlights in Space: Progress in Space Science, Technology, Applications, International Cooperation and Space Law, Vienna, 1997.

and the active US Iridium 33, an incident said to have "unchained a new chapter in the field of space debris".[43]

The 1972 Liability Convention, the treaty that was supposed to address this kind of situation, was unable to address it. Establishing the responsibility for the collision, as the LC provisions require, was impossible because the Russians claimed the satellite was junk and there was nothing they could do to avoid the collision, while Iridium did not even attempt a corrective manoeuvre, considering it too unlikely to happen. As it turned out, low likelihood did not translate into impossibility. In addition, being that Iridium is not a country but a private corporation, it fell outside the remit of the LC, leaving the case in an actual legal vacuum.

A final and still lingering issue was the responsibility for the debris resulting from the impact, which, again, represents a good example of how the existing framework is no longer apt to address and regulate the space sector as it exists today. What is certain is that the impact of Iridium 33-Cosmos 2251 has considerably worsened the debris situation, already critical after the 2007 China ASAT test. According to some models, the two events have produced about 5,500 debris particles, 90 % of which are still in orbit. They account today for 36 % of all LEO objects, threatening to start the already mentioned "Kessler effect" with nefarious consequences for space-related human activities. However, these two events had the positive effect of attracting (again) international attention on the subject and spurring reflection on possible solutions. Specific solutions are available to address the specific issues posed by the debris – collectively known as active debris removal (ADR) systems. One such initiative is "ESA's Clean Space", which looks at ways of cleaning up the space environment as well as preventing new debris from being created in orbit. Creative solutions have been proposed, such as the project, spearheaded by JAXA (Japan's Aerospace Agency), to commission the manufacturing of an electrified 'space-fishing' net that would draw and capture metal debris to eventually burn in the Earth's atmosphere.[44] Alternative ways of dismissal – such as leading end-of-life satellites into a "graveyard orbit", just above the GEO's altitude or into the "spacecraft cemetery" deep down in the southern Pacific Ocean – have been adopted time to time with some success. Of course, any measure involving active measures of mitigation attracts suspicions from the part of the other space players, which can harbour reasonable worries of the adoption of ADRs for harmful purposes.[45]

---

43 Maureen Williams, "Space Debris as a 'Single Item for Discussion'". In *Proceedings of the International Institute of Space Law* (2012), 333.

44 Jim McCurry, "In space, no one can hear you clean". *The Guardian*, 28 February 2014, Sect. 25.

45 Robin Biesbrok, *Active Debris Removal in Space: How to Clean Earth's Environment for Space Debris*. Charlestone: CreateSpace, 2015.

Dual-use concerns notwithstanding, the crucial aspect in the discipline of space debris is still constituted, as the Iridium case highlighted, by the legal liability aspects. The solutions are legally problematic as well, due to some peculiarities of space law. While anyone can size an abandoned vessel in international waters without the ship pavilion or ownership under international maritime law, no one can move a spacecraft belonging to another state in space, complicating any attempt at interstate cooperation and assistance.[46] If anything, a more recent case such as the fall, in August 2018, of Chinese space station Tiangong to Earth, has made the topic of immediate relevance again, even though no harm resulted from Tiangong's fall. But counting on luck should not constitute a basis for policy.

Space as a New 'Tragedy of the Commons'? Scenarios and Solutions

If it is true that the realist approach in space adopts the so-called "high-seas" framework[47], it is also possible that space might be suffering from the so-called 'tragedy of the commons'[48], a scenario in which independent groups make self-interest-driven, rational choices but still end up depleting shared, limited resources, although it is not in anyone's long-term interest for this to happen. History has many examples of this happening. And while a perfect equivalence in space is not possible – as it has been rightly observed, what is unique about space is its character of infinity and of absolute novelty – there are some common points that can be drawn in terms of lessons for the future, together with scenarios. Some of them are optimistic. Moltz designed a few scenarios of space governance, from one state ruling the whole system to a supranational institution of the kind of Star Trek's United Federation of Planets.[49] There are a lot of possibilities in between, but having a weak international organization is worse than having none (everybody remembers the League of Nations of the 1920s and 1930s), and the more crowded the orbits in terms of actors (public, private, even tourist operators) and space objects of different kinds, the more important a regulatory framework becomes. Those scenarios have been described in terms of their consequences on the development of space industries in the conclusion, and they are set on a con-

46 Stuart Clark, "Who you gonna call? Junk busters!" *New Scientist* (2777) (2017): 46–49.
47 Thayer Mahan, *The Influence of Sea Power Over History*.
48 Garrett Hardin, "The Tragedy of the Commons". *Science* 162 (3859) (1968): 1243–1248.
49 James C. Moltz, *Crowded Orbits: Conflict and Cooperation in Space*. New York: Columbia University Press; 2014.

tinuum between conflict and cooperation and between one hegemonic power that leads the global space policy and a cohort of competing powers.

Scholars are divided in their evaluations, though; and if there are some, such as Everett Dolman[50], while some consider the scenarios with one hegemonic power (USA) likely, even plausible, others stress out that the evolution itself of the space industry points in a different direction – e. g., the USA can't protect all its satellites, no matter how much it increases its military expenditures.[51] Another possibility is soft governance, which hypothesizes a set of "piecemeal limited treaties" of the kind proposed by Hertzfeld and von der Dunk[52] and ad hoc solutions case by case, like an International Code of Conduct for Outer Space Activities.

The peculiar characteristics of the medium need to be taken into account. "Orbital dynamics" means that all states share a common border in space, so the actions of one state affect all states.[53] There is, therefore, general consensus about the fact that specific security requirements for space will impose, as much as they did for high-seas, cooperative agreements at an international level.[54]

The reality of a full-fledged war in outer space would be so damaging in terms of consequences for everybody on Earth that preventing it becomes an absolute priority, as it is space defence-building capabilities that are not mistaken for offensive by the others.[55] This is particularly important, given that the "distinctability of defensive from offensive weapons" has been identified in the doctrine as one of the two conditions triggering security dilemma dynamics, with all that follows.[56]

The main hurdle of ensuring international cooperation remains, and it is what has been observed with the UN space treaties, namely, the lack of sanctions and enforcement mechanisms. In an enlightening discussion, Mutschler showed that all mainstream IR approaches have so far fallen short in explaining this lack of co-

---

50 Dolman, Astropolitik.
51 Nancy Gallagher, "The Logic for Space Arms Control". *Bulletin of the Atomic Scientists Roundtable*, 2015.
52 Henry R. Hertzfeld and Frank von der Dunk, "Bringing Space Law into the Commercial World: Property Rights Without Sovereignty". *Space, Cyber, and Telecommunications Law Program Faculty Publications*, 2005. Available online at: https://digitalcommons.unl.edu/spacelaw/15/
53 Townsend, "Strategic Choice and the Orbital Security Dilemma".
54 J. Sumida, "Old Thoughts New Problems: Mahan and the Consideration of Space Power". In C. D. Lutes & P. L. Hayes (eds.), *Toward a Theory of Space Power.* London: Institute for National Strategic Studies, 2015.
55 Bryan Bender and Jacqueline Klimas, "Space war is coming and the US is not ready". *Politico*, 6 April 2018, accessed 22 August 2018. Available online at: https://www.politico.com/story/2018/04/06/outer-space-war-defense-russia-china-463067
56 The other being the evaluation of "whether the defence or the offense has the advantage". See Robert Jervis, "Cooperation under the Security Dilemma", *World Politics* 30, no. 2. (1978): 167–214.

operation witnessed in recent years in space security. The main issue, according to neo-institutionalism, is dominant beliefs about the value of unilateral space policies.[57] If anything, this is probably a telling sign that a new, and more comprehensive, security approach is needed in the space age if a concrete solution in terms of sustainability has to be found.

Conclusions

The debate about space security has just started. As it has been highlighted, "space is simultaneously a powerful enabler for the space-faring nations in military terms but also a critical vulnerability."[58] ASAT testing and debris exemplify this aspect more than well. As it has been demonstrated in this article, the use of ASAT is far more dangerous than it is currently perceived, and the dangers are twofold – both military and commercial. And as more countries join the space club and more satellites are launched into orbit, the situation has the potential to spiral out of control.External and unpredictable factors, such as the unfolding of a war crisis in Europe (the Ukrainian War), can only exacerbate the risks further.

A solution is needed here to make sure the Earth's orbits remain viable. The delivery of critical services from space assets requires the constant availability of space assets themselves. Security in space is also important, both for security from space and for a wide range of non-security applications. According to Pellegrino and Stang, "there is thus an alignment of interests between security communities and the wider public to pursue a space security discourse to preserve the stability and sustainability of the space environment and ensure freedom from threats to the effective access to, and use of, outer space."[59] Therefore, a new regulatory framework is needed to regulate (or prohibit altogether) the use of ASAT. In general, international consensus must be reached for outer space to remain the "province of humanity" to be used for the common benefit, as it has always been intended to be.

---

57 Max Mutschler, *Arms Control in Space: Exploring Conditions for Preventive*. London: Macmillan, 2015.
58 CSIS. Space Threat Assessment. 2018.
59 Massimo Pellegrino and Gerald Stang, *Space Security for Europe*. Paris: EUISS, 2016.

Ipshita Bhattacharya
Chapter VI
Arctic the Zone of Geo-Politics: Risks of Living in Anthropocene

Science has contributed enormous evidence of human interference in Arctic climate change and the plausibility of increased rigor in the future with varied dynamics. The apparent impact is and will be observable at a much more acute level across the world in the environment; climate change, natural resources, vegetation, wildlife, marine life, to name a few. Perhaps the more serious dimension of the Arctic is that it is gradually becoming a powder keg for the global politics between the USA, Russia and China, as they seek to escalate their claims to the natural resources of this region by strategic and military means. Currently, the Arctic is suffering from a speedy systemic transformation with a varied range of economic, climatic, social, political, and security implications that are, however, still vaguely understood. China's Arctic ambitions through its "Belt and Road Initiative" and engagements with Russia in forming new marine routes have stoked US interest in the region. This chapter seeks to explore primarily how the political and security engagements of China, the USA, and Russia are a threat to the region and secondly, what dynamics might occur if they fail to manage and resolve the future challenges of this region.

Risks of Being in Anthropocene

Over the last three decades, the Arctic has rapidly warmed at approximately double the rate as compared to the entire globe. This process is commonly known as "arctic amplification." The principal reason for this phenomenon is largely believed to be man-made climate change. One of the major threats is the floating sea ice cover of the Arctic as it is melting and defrosting rapidly, specifically during the summer. The snow-covered lands in the Arctic are also declining, with permafrost thawing in many various portions. Since the 1980's, the noticeable changes have prominently surfaced.

According to the scientists who closely monitor the region, the changes in the Arctic are worrisome because they could result in further catastrophic deterioration. When ice melts due to global warming, it leaves areas completely exposed to the sun's rays. This water absorbs more heat from the sun, and this additional heat causes the ice to melt further. When permafrost thaws, the flora and fauna that

https://doi.org/10.1515/9783111081687-007

were frozen beneath the ice or between the layers begin to decompose, releasing carbon dioxide and methane into the atmosphere, which speeds up the warming process. Even the changes in Arctic vegetation, which influence the surface brightness, can contribute to warming. As the Arctic atmosphere warms, it becomes more saturated with water vapor, a significant greenhouse gas that contributes more to the warming of the atmosphere.[1]

According to a report by the Intergovernmental Panel on Climate Change (IPCC) of the United Nations, the Arctic ice cap is melting much faster than ice in other parts of the planet. Not only is the release of greenhouse gases leading to Arctic ice melting a serious concern because it is causing changes in the region, but also because it has political ramifications.[2] It is evident that an increase in human activities in the region, such as commercial shipping, oil and gas exploration, and military activities that generate soot emissions from maritime vessels, will exacerbate the region's climate. In the last sixty years, winter temperatures in Alaska and Western Canada have increased by approximately seven degrees Fahrenheit. This increases the region's accessibility to human activity. This implies that the Arctic must make new adjustments in the near future in order to meet human needs.[3] In 2009, two German commercial ships traversed the Northern Sea Route, which was formerly an impenetrable zone, i.e. from Vladivostok to the Netherlands[4] indicating clearly that the region is becoming increasingly accessible.

The opening of these passages, such as the North East and North West, as a result of melting ice, will unleash a Pandora's box of questions concerning the rights of those in control over seaways and the exploitation of vast unexplored reservoirs of natural resources. The greater the human activity and access in the region, the greater the geopolitical concerns will become. Considering the past tensions between the Arctic Coast States and the other players in this region, including NATO, this situation becomes more alarming. These other states may not be geographically close to the region, but they share China's interest in the Arctic's natural resources. In light of what has been said and understood about the

1 National Snow and Ice Data Centre, 2021. Available online at: https://nsidc.org/cryosphere/arctic-meteorology/climate_change.html
2 Horing Spohr, L. G. Cerioli, B. Lersch, and G. Soares, "The Militarization of the Arctic: Political, Economic and Climate Change". *UFRGS Model UN Journal* (2013): 11–70.
3 Scott G. Borgerson, "Arctic Meltdown: The Economic and Security Implications of Global Warming. Foreign Affairs". *Council on Foreign Relations* 87 (502) (2008): 15–19.
4 Heather A. Conley, "A New Security Architecture for the Arctic". *Centre for Strategic and International Studies CSIS*. January 2012. Available online at: https://csis-website-prod.s3.amazonaws.com/s3fs-public/legacy_files/files/publication/120117_Conley_ArcticSecurity_Web.pdf

sensitivity of the Arctic environment, it appears evident that even the slightest disruption of the Arctic environment's homeostasis can result in its warming and severe consequences.

Calculating the consequences of commercialization and militarization of the Arctic in the current global context is frightening. Globalization, technological advancements, and climate change are primarily responsible for the Arctic's complex transformations. This epoch of the Anthropocene has been a crucible of interactions between nature, scientific progress, and the politics of international competition for the Arctic. Over the years, the Arctic has experienced unprecedented physical, economic, and political changes. These rapid changes are causing a significant security threat in the region. The Past recounts the region's decades of peace and stability following the end of the cold war. Perhaps now was the time to shift away from bipolar and military-centric security narratives and toward a more people- and society-centered approach. It is plausible to conclude that regional organizations such as the Arctic Council and the Barents Euro-Arctic Council centered their missions on this broader scope of security issues (BEAC).

The effects of climate change will not only disturb the Arctic region, but will also have other global repercussions. The disruption of the world's climate system and the circulation of the oceans will ultimately result from the rapid melting of ice, which is causing a rise in sea level. This regional impact of climate change is largely the result of human intervention, marine activities such as oil and gas extraction, commercialization of naval route developments, and militarization of the region. This study primarily analyzes and interprets the 'foot prints' of great power politics present in the region in the midst of climate change and even significantly contributing to the crisis. The risk of unwelcome escalation of military tension in the Arctic as a result of deteriorating relations between the United States, China, and Russia over potential disputes in other regions of the world has a greater impact on this region. A consistent increase in the military expenditures of these states and investment in military activity in the region could pose a threat not only to the Arctic environment, but also to global security as a whole.

As a result of the Ottawa Declaration, the Arctic Council was established in 1996 as an intergovernmental organization with the objectives of promoting cooperation, coordination, and interaction. This initiative was intended to foster collaboration between Arctic nations and Arctic residents. The eight circumpolar nations that make up the Arctic Council, namely Canada, Denmark, Iceland, Finland, Russia, Sweden, and Norway, have vowed to protect the Arctic environment and its inhabitants. Two of its members, the United States and Russia, are heavily militarily involved in the region, raising questions and concerns. In other words, it is the first signs of a second Cold War era, as well as the potential consequences if military operations go awry.

Over the years, the ambition to manage the Arctic as a zone of low tension and extensive cooperation appears to have been continually challenged, particularly in the current global political environment. Rising temperature is the primary cause of sea thaw, which in turn promotes commercial activities, which ultimately lead to political leverage and maneuvering among states. The increasing interest and political, commercial, and military interventions of states from outside the region contribute to the region's challenging dynamics. Regardless of whether the Arctic has its own regional governance structure, the constant interferences of states, their competitive political discourse, military deployments, and the spillover that results from this equation may create a situation that leads to threats. Cold War military operations in this region, as well as military buildups populating the region, indicate that the future of the Arctic is precarious. The USA's military presence as retaliation to the Chinese and Russian military buildup in the region can result in a huge crisis. All of these nations are investing heavily in military infrastructure to strengthen their grip on the region, with the United States' participation in NATO (North Atlantic Treaty Organization) introducing multiple new dynamics. In recent years, there has been an unprecedented increase in military tensions between the United States and Russia in the Arctic. NATO's participation in the Barents Sea and Russia's military presence in the Bering Sea are provoking suspicion among the world's leading nations. According to a report by the Arctic Institute, the United States has doubled its military ventures in the Arctic from 2015 to 2020, with the aim of achieving total Arctic dominance. Russia has directed its nuclear arsenal quota up to 81% in its Northern Fleets, and the strategic objective of both the global powers is to possess Arctic dominance in the region.[5]

Considering the current state of affairs and the potential observations that could be made, it can be stated that these potential gaps contribute significantly to the formation of rifts between nations. These gaps can be categorized as governance gaps, such as restricted dialogues and lack of transparency regarding military concerns among nations, limited competence and capacity to execute governance agreements, and tension between the growing need for inclusivity and Arctic interests. However, while these differences do not directly cause disagreements, they do have the potential to incite nations to engage in conflict.Large reservoirs of untapped minerals and other natural resources can spark heated debates among circumpolar nations over the demarcation of their respective economic zones, thereby adding to the potential for discord. The Lomonosov

---

5 Jen Evans, "The History and Future of Arctic States Conflict: The Arctic Institute Conflict Series". *The Arctic Institute*, accessed 25 May 2021. Available online at: https://www.thearcticinstitute.org/the-history-and-future-of-arctic-state-conflict-the-arctic-institute-conflict-series/

Ridge (central portion of the Arctic Ocean) dispute is one such instance that has sparked heated debates and arguments in the political corridors of Canada, Denmark, and Russia, with Russia and Canada claiming the greatest presence on the continental shelf.

The Arctic's enormous mineral wealth, including manganese, gold, copper, iron, and diamonds, as well as its substantial oil and gas reserves, may be the cause of competitions and rivalries between nations. According to the United States Geological Survey, the Arctic contains 25 percent of the world's unexplored oil and natural gas reserves, thereby reserving a significant role in future energy theater. It goes without saying that this reality is a major source of discord among Arctic nations. Due to the Arctic's extreme sensitivity and vulnerability to climate change, any increase in human accessibility in the region will have significant ecological, sociological, political, and economic effects in the region, and these mechanisms will eventually lead to a violent and catastrophic response in the Arctic.

US, China and Russian Intervention: The Geometric Progression of Ramifications

Russia's Arctic Policy

Russia has the longest coastline of all the Arctic States, with the Arctic accounting for approximately 15% of Russia's GDP and 20% of Russia's exports, including 80% and 17% of gas and oil respectively.[6] In addition, the altering security dynamics and their consequential effect on regional geopolitics all contribute to the Arctic's significance to Moscow. It is no longer a hypothesis that Russia views the Arctic's melting ice as an opportunity for its natural resource-based economy, but Russia's capacity and infrastructure expansions in the region are apparent. Arctic is also crucial to Moscow's ambitious goal of achieving a 20% share of the global Liquefied Natural Gas (LNG) market by 2035.[7] As a result of global warming, the Arctic's resources serve another purpose for Russia, namely export to Asia-Pacific nations, which is facilitated by the functional Northern Sea routes. Moscow's efforts to mil-

---

6 Dmitri Trenin, "Russia and China in the Arctic: Cooperation, Competition, and Consequences". *Carnegie Moscow Centre*, accessed 31 March 2020. Available online at: https://carnegie.ru/commentary/81407

7 Bloomberg, "Russia Eyes Greater Energy Dominance with Arctic LNG Push". *Moscow Times*, 8 April 2019. Available online at: https://www.themoscowtimes.com/2019/04/08/russia-eyes-greater-energy-dominance-with-arctic-lng-push-a65140

itarize the Arctic are largely motivated by the region's resources. In 2007, Russian President Vladimir Putin re-prioritized the country's Arctic policies, and now, with ambitious plans, Russia's military establishments in the Arctic are having greater geopolitical implications in terms of their strategic significance. Russian policy therefore focuses primarily on three strategic motives for military configurations in the Arctic.

Due to the presence of other stakeholders in the region, the primary objective is to establish a defensive front line against foreign interventions. Finally, to project its power to deter other power incursions in the Arctic region,[8] one of the reasons for this is that Russia considers NSR to be part of its internal waterways, whereas other international stakeholders consider it to be international sea routes. The international community is developing a strong opinion as they view Russia's approach as Moscow's exclusive control and monitoring of the region. Russia has divided its military capabilities into the East and West sectors of its Arctic region. On the Eastern front, various radar systems, such as Sopka-2, are used to monitor international ships traveling from Asia-Pacific regions through the Bering Strait to the Northern Sea Routes. In order to have functional awareness, Russia is also committed to search and rescue missions and functional radar systems operating in the air and naval domains. Sopka-2 provides all types of meteorological data to travelers passing through NSR. By adhering to all of these operations, the goal of a busy maritime traffic in this zone is achieved, as Russia's military footprint in the zone expands. In the central Arctic region, where the Bastion-P and Pantsir-S1 systems have been installed on the Kotelny and Novaya Zemlya islands, it is possible to observe a significant amount of Russian military activity. These defense establishments are advanced protection layers; the primary reason for constructing such a formidable defense structure is to project its capabilities and dominance in order to maintain control over NATO or American interventions in the region.

The Russian military footprint in the Western Arctic is even more advanced and robust, with full offensive capability, especially in Alexandra Land, which is fully equipped for air, naval, and territorial defense. This system is being constructed to establish a multilayered defense system for maritime operations and air denial systems. The primary objective of this robust defense system is to protect Russia's nuclear arsenals, and the secondary objective is to strengthen Russia's Northern Fleet. In April 2019, in the city of Severodvinsk, located on the Arctic coast of Russia, an accident occurred during a missile engine test, resulting in mul-

---

8 Mathew Melino and Heather A. Conle, "The Ice Curtain: Russia's Arctic Military Presence". *Centre for Strategic and International Studies*, 2021. Available online at: https://www.csis.org/fea tures/ice-curtain-russias-arctic-military-presence

tiple fatalities and a significant increase in nuclear radiation. This situation is alarming because Russia has been actively engaged in nuclear testing and military engagements in this region for many years, despite the fact that this region is geographically remote and, as a result, receives less coverage from mainstream media. The Kursk nuclear submarine explosion in the Barents Sea in 2000 was another instance of a Russian naval disaster that resulted in the death of over a hundred crew members. The nuclear waste dumped by Russia over the years from its Soviet-era submarine reactors and nuclear waste, when its northwestern coast was a center for nuclear testing, is another environmental problem in the Arctic.[9] The presence of nuclear radioactive waste in the Arctic's marine bodies and environment raises grave health concerns for the Arctic ecosystem.

In the Severodvinsk region of the Kola Peninsula, the Northern Fleet command is a projection of Russia's most sophisticated and powerful land, air, and maritime capabilities. The Russian Arctic Brigade has revived its Soviet-era posture by re-establishing its military posts throughout the region with full vigor and planning, including the restoration of its formerly operational airbases, radar stations, and other integrated defense systems. The testing of hypersonic cruise missiles and nuclear-powered underwater drones is the greatest cause for concern; this could be the most detrimental geometric progression of expected repercussions from the militarization of the region. Russia is also equipped with vast icebreaker fleets, which are of great importance to the country as they define its military posture and commercial endeavors. In addition to providing clear passage for military and commercial activities, these icebreaker fleets are frequently equipped with modern warfare systems such as Kalibr cruise missiles, etc. In spite of the fact that a rise in temperature has allowed unfettered access to sea lanes for trade and military deployments, it has also pushed countries such as Russia, the United States, and China into a race for dominance that could result in disastrous and catastrophic challenges. Knowing the facts about climate change and its effects in the Arctic, as well as the evolving cooperation or competition between international players, paves the way for future obstacles.

Moscow has made substantial and consolidated financial and infrastructure investments to fortify and strengthen its active presence, influence, and power projection in the Arctic, regardless of the region's climatic concerns. The greatest cause for concern is Russia's persistent nuclear efforts over the years. Almost 67 percent of Moscow's submarine-based nuclear arsenals operate from the Northern Fleets

---

9 "Sources and Risks of Potential Future Contamination", *Nuclear Wastes in the Arctic.* Washington D.C.: Diane Publishing, 1995, 115–169. Available online at: http://large.stanford.edu/courses/2017/ph241/stevens2/docs/ota-env-632.pdf

of the Kola Peninsula and also the Pacific fleet based in Vladivostok, with nuclear submarines consistently operating in the Arctic region, according to a 2015 report by the Danish Defence College. Since 2017, the 200 Motorized Infantry Brigade of the Russian Armed Forces has been operational in the Arctic, and efforts are underway to upgrade these facilities, particularly the air power, defense, and missile forces, in order to counter any emerging threats from the region. In 2015, the Russian military conducted a uniformed exercise with the participation of 45,000 troops, 15 submarines, and 41 warships. These operations were conducted after repeated incursions into the airspace of Finland, Sweden, and the Baltic, which is an interesting fact. This type of provocative occurrence could result in severe military retaliation and cause a major catastrophe in the region. The Arctic's environment is delicate and does not permit tampering with its natural equilibrium; if human activity is not curtailed and continues at the current rate, it will inevitably result in unprecedented losses for humanity.[10] This kind of provocative incidences could result in serious military retaliations and potentially cause huge disaster in the region. The Arctic's environment is sensitive and does not allow tempering with its natural equilibrium, and if human activity is not limited and continues at the same pace, it will cause unprecedented losses to humankind.

China's Arctic Policy

Arctic issues do not play a significant role in China's foreign policy, but its increasing political and economic interest in the region over the past few years demonstrates its strategic objectives for the region. Moreover, China wishes to participate as one of the region's stakeholders. China chose to invest heavily in the Arctic to determine its economic, diplomatic, and security goals. Polar Regions were thus included in China's twelfth Five Year Plan in 2011, followed by China's Arctic Policy in 2018 and the inclusion of Polar Silk Road in the One Belt One Road (OBOR) program, making Beijing's commercial and security goals in the region clear.[11] China's main goal in the Arctic is primarily commercial at present, but it is increasing its diplomatic and scientific efforts most strategically to support its primary agenda. The international community is becoming increasingly concerned about China's in-

---

10 Bert Chapman, "The Arctic's Emerging Geo-politics: Recommendations for the US and its NATO Allies". *Mackinder Forum*, 12 October 2020. Available online at: https://mackinderforum.org/the-arctics-emerging-geopolitics-recommendations-for-the-u-s-and-its-nato-allies/
11 Heljar Havnes and Johan M. Seland, "The Increasing Security Focus in china's Arctic Policy". *The Arctic Institute*, 16 July 2019. Available online at: https://www.thearcticinstitute.org/increasing-security-focus-china-arctic-policy/

volvement in the region, which becomes even more concerning when compared to the militarization of the Arctic by Russia. The United States Coast Guard (USCG) has reported that China and Russia could pose a threat to the global rule-based international order, primarily due to China's similar behavior in the South China Sea and East China Sea. Beijing's active policies of infringement in Asia and its aggressive stance in no way contribute to its benign Arctic involvement.

China is unquestionably emerging as a key player in the Arctic, and as a result, concerns regarding China's policies and actions have assumed a prominent position. Notable is China's transformation over the years from an isolated partner to an active member of the Arctic Council interpreting its sole strategic objectives. China published a white paper on its Arctic policy in 2018, outlining its intentions and goals. Positioning itself as a 'near Arctic State,' China pledged to actively participate in the region's issues. China reveals its greatest concerns regarding climate change, which have direct effects on its economy. China's entry does not come as a surprise, given that it is the world's largest energy consumer and will therefore have the most vested interest in the region. It is also understood and clearly perceived that the country is eager to use and operate via maritime routes to extract hydrocarbons from the Arctic. Beijing intends to connect the Belt and Road initiative to the 'Polar Silk Road,' as detailed in its white paper. China intends to construct the Polar Silk Road through the Arctic in order to connect Asia and Europe and facilitate the opening and operation of logistics and transportation channels. This implies a commercial involvement in the region that is so intense that it threatens the region's natural equilibrium.

China has developed a two-pronged strategy for the Arctic, focusing first on scientific research and surveys, new avenues for resources and commercial purposes, and secondly, on determining the root cause of climate change. As it is now a well-known fact that the thawing of ice and melting of the seas is creating a new regional order that will result in new statecraft and real politics in the Arctic zone, Beijing may be attempting to justify its presence and also pretend to contribute to the cause in this manner. The white paper emphasizes the objective of the "Polar Silk Road," which is the continuation of the Belt and Road Initiative, which is to have well-developed and functional ports and transportation routes. According to this plan, the route will pass through several strategic points, including the Bering Sea, the Northern Sea Route, and the Norwegian Sea.

China is circumnavigating Asia and Europe and entering Africa primarily via three major configurations: the Maritime Silk Road, the Belt and Road, and the Polar Silk Road. China's unwavering aspiration to ascend to the world's preeminent position motivates its intense involvement in Arctic affairs. China achieves this objective through aggressive policies. no longer wishes to be confined to Asia as a developing nation, but rather wishes to expand its role as an agenda-setter. The Silk

Road is a strategic means for China to demonstrate its power, influence, and dominance in order to fulfill its national interests, which are supported by economic and military factors. Analyzing China's Arctic policy reveals that China is carving out a dominant, commercial, and strategic identity for itself in the Arctic. It has articulated its economic interests while simultaneously projecting its dominant position as a polar power player. China is compelled to make substantial investments in the region by commercial forces, security concerns, and the allure of abundant energy supply chains. As this will have a significant impact on the Chinese energy sector and its monitoring capabilities, Chinese energy firms are keeping a close eye on the region's oil and gas reservoirs and competing for the same. In the context of the Arctic, it is common knowledge that climate change has created new opportunities for the commercialization of the region, and in order to dominate the commercialization, stakeholders and other international players are militarizing the region. However, this is creating a new geopolitical conflict theater for states to engage in fierce rivalry. However, China's future stance will determine what new dynamics may emerge in the region, particularly when other actors are also present.

The assertive rise of China and its encroachment on the position of superpower will inevitably result in a contested environment in the Arctic and its coastal states. In an environment of renewed tension between the United States and Russia and the United States and China, the Arctic is undergoing a rapid transformation; China's newfound presence in the region is viewed as a permanent threat, as Beijing is engaged in similar military and commercial endeavors in other regions of the world. However, amidst these geopolitical struggles, competitions, and rivalries, the major international players appear oblivious to the impending environmental repercussions. It appears that the region is simultaneously facing geopolitical rivalries, competitions, and contests; Russia has also raised concerns about the United States and NATO's involvement in the region, while Moscow acts in accordance with China for their collective interests.

USA's Arctic Policy

In the Arctic, the effects of climate change have begun to manifest as a reduction in ice, an increase in human access, and the economic exploitation of the region. The extent of these effects will depend on how quickly international powers grasp this reality. The United States and its allies are pursuing internal configurations that will deepen their mutual collaborations. Due to the assertive stance jointly adopted by Russia and China, the United States deems this collaboration to be essential. The alignment of the United States with NATO could push Russia and China toward each other in a more dynamic manner, adding an additional dimension to this arrangement. The United States' objectives in this region are to defend and collaborate with allies and partners to advance shared interests. The United States has proposed establishing a secure environment in the Arctic in order to reduce threat perceptions, enhance functional capabilities, and establish and support a rules-based order.

This year, the US Department of Army unveiled its Arctic policy, which is primarily aimed at regaining dominance in the Arctic by deploying an army capable of overcoming all obstacles. Major emphasis has been placed on its strategic posture, training, logistics, ability to fight and survival in the harsh Arctic climate. In order to accomplish this, the US forces have established a two-star operational headquarters with all kinds of advanced amenities to combat high altitude challenges. According to the US Department of Defense's Arctic strategy, titled "Regaining Arctic Dominance," it includes essential requirements for the army's training, organization, and equipping with the latest and most advanced equipment suited to the environment, as well as collaboration with other Arctic allies. These strategies are implemented to protect US interests and preserve regional peace and tranquility. In order to justify the placement of NATO forces in the Arctic, the collaborative efforts with allies are given a great deal of importance. However, the release of this strategy can be viewed as timely, especially given the rise of great power politics in the region, whose militaries' powerful projections only serve to increase competition and contests.

The primary concern for the Arctic region is the expansion of military activities and the resulting escalation of tensions; recent developments are increasing the likelihood of a military clash between the United States and Russia due to the involvement of NATO and Russian vessels in the Barents and Norwegian Seas. The important NATO military exercise 'Trident Juncture' held in Norway in 2018 sparked controversy because the Russian military jammed GPS communication services. In addition, similar situations may escalate in the future because Russia has bases on the Kola Peninsula and NATO has bases in Norway, which are separated by only a few kilometers. Seven Arctic nations, excluding Russia, include five NATO members (the United States, Denmark, Canada, Iceland, and Norway) and two NATO partners (Finland and Sweden). As a result, the decades-long rivalry between Moscow and Washington will continue to raise questions and challenges regarding the stability and future of the Arctic. The abundant availability of energy resources is an extremely enticing and motivating factor that keeps the Arctic at the forefront of this region's geopolitics.

Climate change and technological advancements in the extraction of natural resources are significant contributors to the engagement of states in this region. The Arctic contains approximately 90 billion barrels of oil, 1,669 trillion cubic feet of natural gas, and nearly 44 billion barrels of natural gas, according to a

USGS geological survey, with approximately 84 percent of these resources located offshore. This information also suggests that approximately 70 percent of the offshore oil reserves exist in the Arctic's Alaska, Greenland, and Barents Basins, and that almost 70 percent of the unexplored natural gas reservoirs exist in the West Siberian Basin, Arctic's Alaska, and Barents Basin. This data explains the cause of geopolitical turbulence that is assuming an ugly form, provoking military escalation and causing grave global concerns.

Over the years, the Arctic is gradually becoming a high-alert zone and receiving some unsettling wake-up calls regarding the region's advanced militarization. This region, which was once known for its aversion to conventional hard politics, appears to be defying the situation and the prevailing conditions, rather aggravating the return of military conflicts. The gradual introduction of the military into the region by Arctic players for their respective agendas may have misconstrued the region's security. When the term Arctic amplification is understood in its real-world context, provocations and counter provocations, military and nuclear activities in the Arctic appear too dangerous.

Dynamics of the Climate Change

Temperature increases are causing a dramatic decline in Arctic ice, including sea ice, and accelerating the thawing of permafrost. The Arctic has a significant impact on global climate, and consequently, a significant reduction in Arctic ice is affecting global weather patterns. The global impact of climate change, which results in droughts and floods exacerbated by global warming, causes food and water shortages. This further contributes to mass migration, which destabilizes the demographic equilibrium and can even result in migration conflicts. We know that the Arctic temperature has played a crucial role in maintaining the stability of the global climate. Now, however, as a result of human-caused global warming, the temperature is rising at an alarming rate, resulting in Arctic amplification. Globally, the average temperature has risen by nearly one degree Celsius, but in the Arctic, it has increased by approximately two degrees Celsius over the same time period.[12] The rapid warming of the Arctic ice is causing a massive loss of sea ice, and by the end of the summer of 2019, the area of sea ice in the Arctic was recorded as the second lowest since 1979.

---

12 Benjamin J. Sacks, Scott R. Stephenson, Stephanie Pezard, Abbie Tingstad and Camilla Sorensen, "Exploring Gaps in Arctic Governance". *RAND Corporation*, 22 September 2021. Available online at: https://www.rand.org/pubs/research_reports/RRA1007-1.html

According to various scientific global warming simulations regarding the region's future forecast, it is estimated that the Arctic Ocean will be completely devoid of ice for the first time before 2050. According to Peter Wadhams, the global warming caused by the land-snow-ice feedback is roughly equivalent to the additional 25 percent of global CO2 emissions that are responsible for the disruption of the Arctic's atmospheric behavior and oceanic circulation patterns.[13] As the Arctic Sea ice continues to melt due to global warming, an increasing number of individuals are gaining access to the region and engaging in activities there. This is an undeniable truth that appears to be becoming more evident daily. Unfortunately, human access and military involvement in the region have transformed the warming of the Arctic from a consequence to a cause of global warming. It has been observed that the loss of Arctic ice is directly related to the extreme climate in the Northern Hemisphere, with polar temperatures rising twice as quickly as subtropical temperatures. This change in the atmosphere weakens the jet streams and maintains the same weather pattern for days. In 2018, extreme weather phenomena were caused by a weakening of the polar jet stream. The repercussions of climatic change are visible across the globe, and scientific evidence predicts and demonstrates that larger catastrophes are imminent if harmful human activities are not controlled or monitored.

Unfortunately, human access and military involvement in the region have transformed the warming of the Arctic from a result to a cause of global warming. It has been observed that the loss of Arctic ice is directly related to the extreme climate in the Northern Hemisphere, as the Polar region is warming twice as quickly as the subtropics. This change in the atmosphere weakens the jet streams and maintains the same weather pattern for several days. In 2018, extreme weather events were caused by a weakening of the Polar jet stream. The repercussions of climatic changes are visible across the globe, and scientific evidence predicts and demonstrates that larger catastrophes are imminent if harmful human activities are not monitored or controlled. If substantial loss of Arctic ice's reflective capacity persists, it will result in a warming equal to one trillion tons of CO2 and will accelerate the two degree Celsius threshold by twenty-five years.[14] Consequently, world leaders should be responsible for any pragmatism aimed at preventing this. This sensitive region necessitates adherence to natural norms and respect for nature, as the warming of the Arctic will have direct or indirect global repercussions. Understanding the significance of the Arctic necessitates the investigation of potential

13 Peter Wadhams, "Arctic ice cover, ice thickness and tipping points". *Ambio* 41 (2012): 23–33.
14 Meinrat O. Andreae and V. Ramanathan, "Climate's dark forcings". *Science* 340.6130 (2013): 280–281.

disasters if the Arctic is not managed properly. According to research and surveys conducted on the region, it is alarming to learn that if human access and military interventions continue, the Arctic may vanish entirely by 2050. The Arctic is currently covered by sea ice throughout the year; however, the ice decreases during the summer and grows back during the winter; however, it is possible that the area covered by ice has decreased at a rapid rate over the past few decades.

Various research institutes and simulation models predict that ice will continue to decrease due to global warming, which will have a significant impact on the Arctic ecosystem and climate. As the Arctic sea ice is melting and decreasing, there has been a significant variation recorded during this transitional period over the years, which has a significant impact on the lives of the local people and other ice-dependent species. According to a recent study, the evolution of Polar Sea ice cover under conditions of high CO_2 emissions and inadequate climate protection is predicted by forty distinct climate simulation models. As anticipated, the summer Polar Sea ice disappeared rapidly in these tested CMIP6 simulations; they also observed that ice disappeared rapidly in simulations where CO_2 emission was rapidly reduced.[15] The Arctic is warming at a rate that is twice as fast as the melting of ice in the rest of the world. This is because the ice, sea, land, and atmosphere of the Arctic are so interconnected that any changes in one component will directly impact the others and also trigger the entire system. This Arctic amplification is more apparent during the winter, when it is also a major concern that the Arctic is receiving less snow and forming less ice. According to a report published in 1979, the Arctic Ocean has lost nearly 30 percent of its ice cap. Moreover, the ice that exists today is extremely thin and fragile, making it easily mobile.

All of these climatic effects have severe consequences for human, marine, and animal ecosystems. According to the 2019 World Ocean Review, due to the warming of permafrost, a large portion of the Alaskan and Siberian coastlines are rapidly eroding, and the once-frozen land surface is thinning and losing its tenacity, causing collateral damage. Similarly, according to the Norwegian Polar Institute 2020 report, the current and anticipated climate changes in the Arctic region over the next few decades will continue to impact and interfere with atmospheric circulation, vegetation, and most importantly the carbon cycle, and thus have a significant effect on the overall climate system within and beyond the Arctic. Due to the compounds' reactivity to organisms, which is temperature-dependent, climate change will also significantly impact the transport and distribution of eco-toxins.

---

15 Dirk Notz, "Geo-Physical Research Letter". *AGU Advancing Earth and Space Science*, published 16 May 2020. Available online at: https://agupubs.onlinelibrary.wiley.com/doi/epdf/10.1029/2019GL086749

Climate change will have enormous repercussions due to the melting of sea ice, which in turn will impact the radiation balance in the overall climate system through the albedo effect, wherein warming will directly influence the water formation in the bottom through the surface warming of the sea, and this will also have an effect on the "motor" in the ocean system, which in turn will determine the structure of the world's overall climate. The melting of glaciers will raise sea levels, and the thawing of permafrost will increase the levels of greenhouse gases, especially methane. Ecosystems gradually adapt to a specific region's climate; as a result, climate change will have a significant impact on the ecosystem, and different ecosystems will react differently to this change. It is also important to recognize that the effects of climate change are intertwined with other factors, such as pollution, changes in land surface and its use, population and human activities, and the economy. Together, these factors can have a multiplied impact on health, well-being, and respective ecosystems.

Conclusion

This chapter has examined the current state and predicament of climate change in the Arctic region, as well as its relationship to geopolitics. As climate change creates more economic opportunities, the study reveals how the world's major powers are attempting to expand their presence and compete for resources in the region. Nevertheless, it is well known that economic activity will enhance commercial objectives while accelerating environmental degradation. The chapter also discussed the military options adopted by the superpowers in order to ensure regional security. However, this also encourages military collaborations, strategic alliances, and engagements that may lead to destructive combat progressions in the region, resulting in severe harm to humankind and the environment.

People inside and outside the Polar regions are impacted by climate change in the Arctic in two ways: first, physical and ecological changes in the Arctic zone have a global socioeconomic impact; and second, physical changes in the Arctic influence all of the significant processes that contribute to global climate change and rising sea levels. Humans, their societies, and their access to natural resources, as well as their depletion or extinction, pose the greatest threat. For example, fishing is a regional and global food source that is vital to the economies of many states, but climate change is altering marine habitats and thereby impacting the ecosystem. There are numerous examples of climate change causing disturbances in ecosystems, but this is largely dependent on human activities and their responses to the extremely fragile and delicate balance of nature.

Building new shipping corridors in the Arctic for cost savings, commercial purposes, and primarily for political implications relating to global trade will not only increase environmental risks in the Arctic, but will also have military and security implications for the region. Since thinning Arctic Sea ice increases human access to offshore petroleum and encourages source extraction, Arctic states and other states in the region are increasingly engaged in competition for resources. This activity is not as straightforward as it may appear; it will have severe repercussions as the race for natural resources is resulting in the establishment and deployment of respective militaries in the Arctic, making it a geopolitically complex situation.

Ben Jack Nash
Chapter VII
What is the Matter with Catastrophes? The Response from an Artist

This chapter presents a series of photos derived from a performative art installation of "What is the Matter with Catastrophes?". A central element of the work revolves around a time-lapse film that depicts a chicken embryo in its early stages of
development. The film begins with the yolk – an abstract, sun-like, yellow monochrome image, which starts to morph and grow. From a network of red capillaries
to the beating heart, we witness the embryo forming into a fully-fledged fetus complete with eyes and beak. The film ends abruptly, uncertain of the fetus' fate before
looping back to the beginning.

The screen displaying the film is made from paper. Rather than acting as a
mere passive support or functional surface, attention is drawn to the screen as
an object, endowed with an active and transformative role. At various moments,
the artist sprays it with jets of water. This causes the screen to gradually dissolve
onto the glass window on which it is mounted. By the end, the screen has significantly disintegrated, with some fragments remaining and the rest dripping down
the surface of the window like rain and revealing parts of the artist's studio behind.

The Greek philosopher and mathematician, Anaximander, articulated how all
physical matter typically begins in an intangible primordial state before becoming
more material and identifiable. The transformations taking place in the film, as on
the screen, speak to a relationship between abstract and material states. A relationship which, in the artist's view, lies at the heart of many of the catastrophes
we have been witnessing.

Whilst the embryo moves from the abstract to material, we see the reverse taking place with regards to the screen's disintegration, an object in the throws of
transition and on the threshold between states of consciousness. The screen begins
as a functional and instantly recognizable object, belonging quite clearly in a material state before breaking down into random, more abstract sludge.

The work represents a technique and process which are being increasingly
used by the artist to compliment metaphorically his development of the "Triple
A" theory. Rather than being grounded in traditional research methodology, Triple
A comes out of the art studio through explorations and manipulations of matter. It
can be said to apply to a large number of socio-political phenomena and is considered no less pertinent concerning observations of the frequency of human-induced

https://doi.org/10.1515/9783111081687-008

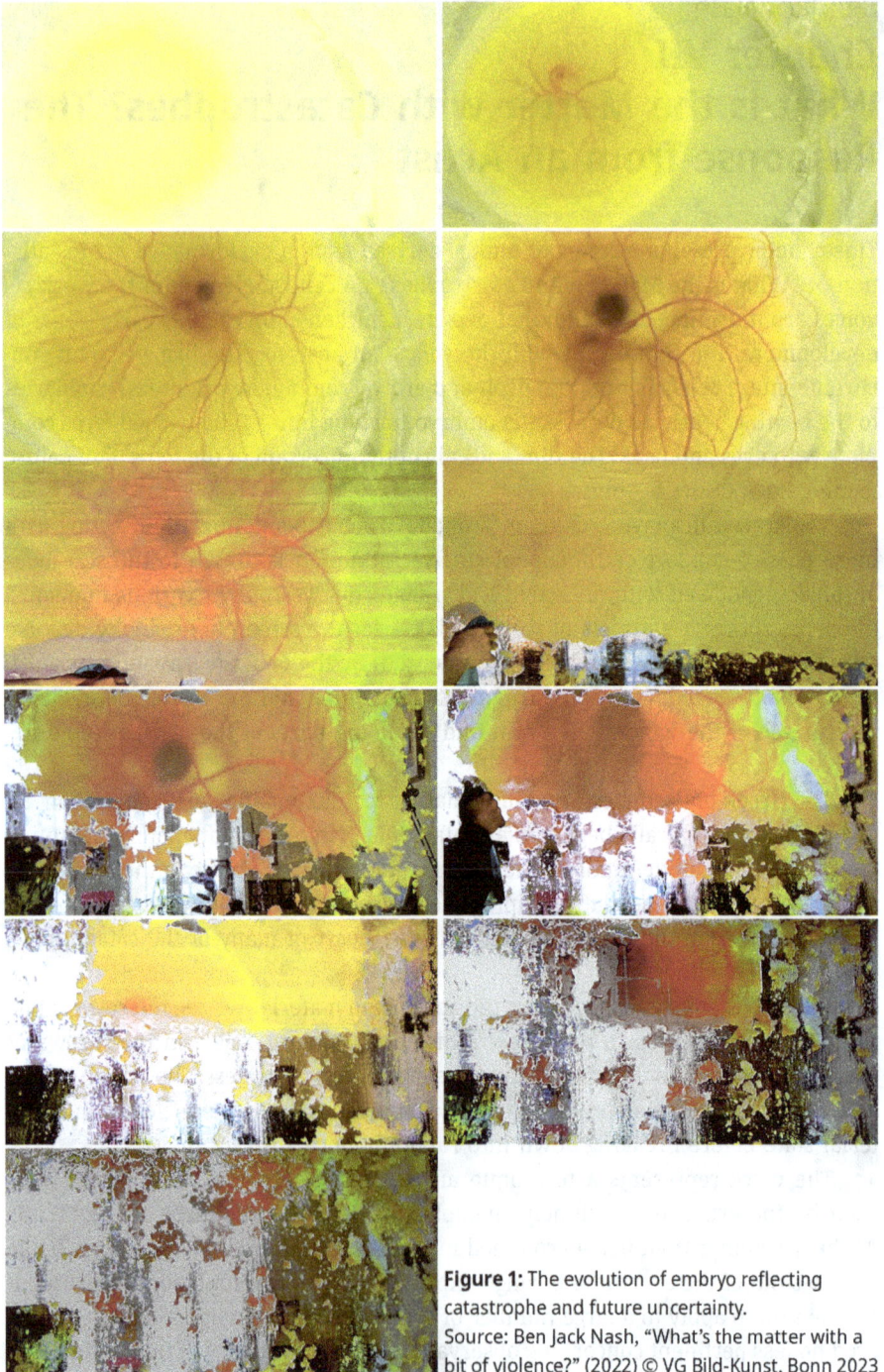

Figure 1: The evolution of embryo reflecting catastrophe and future uncertainty.
Source: Ben Jack Nash, "What's the matter with a bit of violence?" (2022) © VG Bild-Kunst, Bonn 2023

catastrophes. Triple A (or perhaps aaagh!!!) stands for artificially accelerated abstractification. It boils down to how a large number of fundamental and influential processes have been moving towards a more abstract identity: economic policies, political trends, how we communicate and relate to the world, and the state of the planet.

An abstract form can be appreciated by its greater sense of movement, less defined by limits and taking place discreetly behind the scenes. An object in its abstract state radically changes its relationship with space, with movement, and, of course, with time. Its internal structure is less constricted. Matter in its abstract form is unfamiliar, unpredictable, and difficult to empirically measure and reproduce artificially.

Fundamentally, an abstract process is one which can be defined by a far greater interaction with external and environmental elements compared with its material counterpart. Any object can undergo a process of abstractification (sic), but rather than simply using the form and aesthetic to define it as an abstract state, AAA looks far deeper into the process to provide some insight as to the spatial relationships taking place that are responsible for this change of state. One of the key elements is a shift in the objects' relationship from within itself to interact more meaningfully with outside stimuli. At its extreme, it will interact to such an extent that its previous state can no longer be said to have the same identity as it morphs seamlessly with its environment to become part of it.

Abstractification is not itself catastrophic. On the contrary, it is a perfectly natural and, in fact, necessary process that takes place around us all the time, lying at the heart of the universe's existence ad infinitum. The problem arises, however, as we witness time and again, when this takes place at such an unnaturally fast rate that we and the biosphere are simply unable to evolve fast enough. And it is the consequences of this that can pave the way for catastrophic situations to develop.

Climate change is a good illustration of AAA taking place within the physical landscape. When you look at the consequences of climate change on the natural world, what you see is AAA taking place on a macro scale. Melting ice, land erosion, species extinction, bleached coral, forest fires, desertification. Such processes of evaporation, of combustion, of liquifying and so on, all represent physical matter in its more solid and functional form, dramatically shifting towards more abstract states and far faster than normal rates of evolution brought about artificially through human activity. However, AAA extends just as equally to non-physical landscapes such as values and ideology. The nature of politics is similar to the nature of matter and objects, not only in terms of its physical manifestations but also the nature of its characteristics. Culture, values, and social attitudes are also defined by their relationship with space, movement, and time. They can equally be broken down, fragmented, and reconstructed. We use the language of movement and time to label values, for example, as conservative or liberal. Conservatism conjures a desire to hold onto traditions and values – a stagnation of time. And liberalism, very much a spatial term, conjures greater movement and loosening up, or freedom from restraint. And here we can see how neo-liberalism, with its opening up, for example with regard to acceptable social norms, reflects a distinctly more abstract character. This extends equally to the philosophies of free-market economics and gender fluidity.

It is important to recognize the consequences when tectonic shifts in matter and space-time relationships take place. These will continue to lead to instability, as demonstrated by populist and nationalist leaders, insecure ethnic majority populations, and, of course, Mother Earth.

Matteo Nicolini

Chapter VIII
Law and the Humanities in a Time of Climate Change

> Those who remain in a region ravaged by extreme weather often find themselves navigating
> an entirely new social and political structure, if one endures at all.
>
> Wallace-Wells, *The Uninhabitable Earth*, 127.

'A New Subject for the Law' in a 'Human-Altered World'

As members of political communities, we are experiencing how alarming the threats posed by climate change are. Because of our living in the Global North, we thought that 'wealth was a shield against [its] ravages',[1] as well as a 'safe normative space' and a 'safe oasis', within which the disturbing arguments prompted by climate change could be neutralized and brought 'to an end' by consumeristic lifestyles.[2] We have carefully cultivated the illusion that it is possible to save the planet by perpetuating 'the logics of global capitalism and market ideology'[3] and reconciling our ecological commitments with the interests of our unsustainable economic system. Consequently, we kept on 'mortgaging the ecological future of the planet' and, at the same time, allowed its 'cruelest impacts' to affect the less-developed countries (*UE*, 54): 'the systematic pillaging of natural resources, environmental catastrophes ... and endless wars' will indeed be the cost the Global South is most likely to pay for climate change.[4]

---

1 David Wallace-Wells, *The Uninhabitable Earth. A Story of the Future.* (London: Allen Lane, 2019), 1, 54. Further references in the text, abbreviated as *UE*.
2 Respectively, Richard Mullender, "There is No Such Thing as a Safe Space". *The Modern Law Review* 82–83 (2019): 549–576, 551. Stanley E. Fish, *Winning Arguments: What Works and Doesn't Work in Politics, the Bedroom, the Courtroom, and the Classroom.* (New York: Harper Collins, 2017), 3.
3 Sarah McFarland Taylor, *Ecopiety. Green Media and the Dilemma of Environmental Virtue* (New York: New York UP, 2019), 5 and 7.
4 Mikhail Xifaras, "The *Global Turn* in Legal Theory". *Canadian Journal of Law & Jurisprudence* 29.1 (2016): 215–243, 219.

https://doi.org/10.1515/9783111081687-009

On the brink of the final judgement, we are brooding over our acting 'selfishly' (*UE*, 10)[5]. After building 'our way out of nature,' we now realize 'how hard, and how indiscriminately,' climate change 'is hitting' (*UE*, 71, 74). Playing with book titles: climate change undermines both the 'Wealth' and the 'Health of Nations,' causing 'Famines, Fevers,' and extreme weather whose impact on 'the Fate of Populations' results in global mobilization,[6] unleashing as many as a billion migrants by 2050' (*UE*, 133). In this 'human-altered world,'[7] legal scholars witness the disruption (if not the demise) of well-established taxonomies, on which communities have been building their political bonds for the last thousand years.

As a field of scientific research, climate change is not a new topic; climatologists, geographers, political scientists, and sociologists have been studying it for at least 40 years. This has triggered a vast literary production, which explores the causes, manifestations, and consequences of global warming for both the earth and humankind as a whole. Likewise, there is a broad scientific consensus on climate change; although segments of society are skeptical about it, the consensus also extends to its anthropogenic origins.[8] However, climate change is a relatively 'new subject for the law.'[9] The interest in it has emerged only very recently, often intertwining with other legal sub-disciplines, such as international law, environmental law, global law, and migration law.[10] This adds a flavor of cross-disciplinar-

---

5 See *Letters to the Earth. Writing to a Planet in Crisis*, ed. Emma Thompson (London: Collins, 2019), 36. Further references, abbreviated as *LE*.
6 Adam Smith, *An Inquiry into the Nature and Causes of the Wealth of nations*, ed. Kathryn Sutherland (Oxford: OUP, 1993). Anthony McMichael, *Climate Change and the Health of Nations. Famines, Fevers, and the Fate of Populations.* (Oxford: OUP, 2017).
7 Chris D. Thomas, *Inheritors of the Earth. How Nature is Thriving in an Age of Extinction.* (London: Allen Lane, 2017), 29. Further references in the text, abbreviated as *IE*.
8 See e.g. Mike Hulme, *Why We Disagree about Climate Change: Understanding Controversy, Inaction and Opportunity* (Cambridge: Cambridge University Press, 2009). Lorraine Whitmarsh, "Scepticism and uncertainty about climate change: Dimensions, determinants and change over time", *Global Environmental Change* 21.2 (2011): 690–700. Willelm Van Rensburg, "Climate Change Scepticism: A Conceptual Re-Evaluation", *SAGE Open*. April 2015. doi:10.1177/2158244015579723 (accessed 18 July 2022).
9 Anne-Sophie Novel, "Climate change: A new subject for the law", *The UNESCO Courier*, 3 (2019): 13–15.
10 See also Jan McDonald, "The Role of Law in Adapting to Climate Change", *WIREs Climate Change*, 2.2 (2011): 283–295. On the impact of climate change on discrete legal disciplines see e.g. Oliver C. Ruppel et al. (eds.), *Climate Change: International Law and Global Governance*, vols. I and II (Baden Baden: Nomos, 2013). Daniel A. Farber and Marjan Peeters (eds.), *Climate Change Law.* (Edward Elgar: Cheltenham and Northampton, MC, 2016). Benoît Mayer and François Crépeau (eds.), *Research Handbook on Climate Change, Migration and the Law.* (Edward Elgar: Cheltenham and Northampton, MC, 2017). Cinnamon P. Carlarne et al. (eds.), *The Oxford Handbook of Interna-*

ity and supports undertaking innovative research in this area.[11] Disciplinary interactions like these permit probing the potentials of comparative-law methodology in one of the less explored ambits of climate-change-related studies, i.e. global warming acting as a non-legal variable triggering transformations in the legal spectrum. As is evident, climate change raises new issues and requires innovative responses, which will unavoidably affect the realm of the law, which must adapt to environmental changes.

This chapter addresses global warming and climate change-related migration connections by adopting a cross-disciplinary approach. In order to capture the interrelations between climate change, mobility, and the law, it resorts to a non-fictional literary genre (i.e., climate change pop-science). The critical legal approach is complemented by the legal theoretical perspective, which examines how climate change affects the bonds of political communities and the legitimacy of political authorities. It also explores how a strategic use of the law shapes our approach to global insecurities related to migration and climate change. Against this background, the essay maintains that there are also possibilities. The non-fictional texts reflect the ideas of the most active forces within society, fueling dynamism when tackling the ecological crisis. In a time of climate change, these forces stir formalism and make the law act as a bridge, linking our troubled reality to an inclusive future.

Climate Change and Mobility: Perspectives from Law and the Humanities

As stated in the introductory section, the story of climate change has already been told, and I do not intend to repeat it. My purpose is limited in scope: I shall reconsider some common assumptions we usually make about the law, by locating them *in a time of climate change*. A methodological change is required if we want to address 'how our greenhouse-gas emissions warm the climate everywhere' (*IE*, 5).

---

tional *Climate Change Law.* (Oxford: OUP, 2016). Daniel Bodansky et al., *International Climate Change Law.* (Oxford: OUP, 2017). Benoît Mayer, *The International Law on Climate Change.* (Cambridge: Cambridge University Press, 2018).
11 See Erkki J. Hollo et al. (eds.), *Climate Change and the Law.* (Dordrecht: Springer, 2013). Michael Mehling, "The Comparative Law of Climate Change: A Research Agenda", *Review of European, Comparative and International Environmental Law,* 24.3 (2015), 341–352. Matteo Fermeglia, "Comparative Law and Climate Change", in Francesca Fiorentini and Marta Infantino (eds.), *Mentoring Comparative Lawyers: Methods, Times, and Places.* Liber Discipulorum Mauro Bussani. (Cham: Springer, 2020), 237–259.

The cross-disciplinary ambitions of comparative law permit us to grasp the truest politico-legal implications of the present ecological crisis. And this chapter assumes that it is highly productive *to do comparative law in a time of climate change.*

Climate change also challenges our legal paradigms and alters how societies perceive their political bonds: global warming requires what Geoff Mann and Joel Wainwright term the 'adaptation of the political,' because 'it will have massive impact on the way human life on Earth is organized.'[12] Its effects 'have already eaten into trust in state authority ... igniting a complex bundle of social kindling' (*UE*, 128). It then raises concerns about the legitimacy of state policies in confronting ecological emergencies. In an age, like ours, the very idea of borders is being undermined. In transcending them, displaced refugees and mass migration add fuel to the transnational effects of globalization. Likewise, climate change has spill-over effects triggering transformative changes within national orders.

Our troubled times suggest we address how global warming refashions the law by a change of mood. The change is both substantive – i.e. we must adapt the law, its contents and role to existing environmental challenges – and methodological. The paper contributes to this conversation, goes above and beyond the current formalistic approach to climate change, and engages in cross-disciplinary investigations to 'pluralize the debate' in legal research. Conceiving of, experimenting, and deploying innovative methodologies is crucial as regards the 'political, ethical, legal and cultural dimensions of the relation between climate change and migration' is therefore highly productive.[13] Among lawyers, Irus Braverman has shown us how to tackle with cross-disciplinary climate-change related research. In order to understand how global warming affects the 'political' communities of the seas which are coral reefs, she has undertaken 'a massive research ... that stretched across continents and disciplines' with coral experts, whom she considers 'the vanguard of conservation in the Anthropocene.'[14]

The chapter contributes to this conversation adopting the cross-disciplinary stance of law and the humanities, which provides scholars with new 'possibilities'

---

12 Geoff Mann and Joel Wainwright, *Climate Leviathan: A Political Theory of Our Planetary Future.* (London and New York: Verso, 2018), x-xi. Further references in the text, abbreviated as *CL*.
13 Andrew Baldwin, "Pluralising climate change and migration: an argument in favour of open futures". *Geography Compass* 8 (2014): 516–528, 516. On the pluralisation of the methodological debate in legal studies see Jaakko Husa, *Advanced Introduction to Law and Globalisation.* (Cheltenham and Northampton: Edward Elgar, 2018), 165. Jaakko Husa, *Interdisciplinary Comparative Law. Rubbing Shoulders with the Neighbours or Standing Alone in a Crowd.* (Cheltenham and Northampton, MS: Edward Elgar, 2022.
14 Irus Braverman, *Coral Whisperers. Scientist on the Brink.* (Oakland, CA: University of California Press, 2018), 2–3.

and renovated 'perspectives' to assess how climate change affects the legal spectrum.[15] Evidently, my cross-disciplinary legal approach will be that of a comparative legal scholar, whose critical attitude aims to detect how global warming affects the realm of the law and global insecurities.

In a time of climate change, the two constitutive parts of 'law and humanities' play different, albeit interlocked, roles. I have resolved to focus on a non-fictional literary genre, i.e. the climate-change pop-science. Prompted by the insurgent ecological crisis, it comprises essays, pamphlets, letters, and other writings related to the outcomes of our burning the planet, which convey a convincing "callout to the public … in response to climate and ecological emergency" (*LE*, 6).[16]

This genre assists legal scholars in evaluating the socio-legal consequences of how global warming is altering the organizing themes within political communities. To this end, the critical comparative approach will be complemented with the legal theoretical perspective. Changes in how political bonds are now arranged may indeed be properly understood within the context of 'contractarianism,' according to which 'political society is a form of contract produced by the consent of the people.' As 'climate migration is both international and domestic,' our political communities have become 'transnational, communities whose common concern is how climate change affects our lives. The phenomenon is transnational because it disregards the role territory has traditionally played in organizing communities – and, despite this, it has ostensible effects on both communities and territories.

In the following section, I first consider how climate change affects the bonds of political communities and reflect on how it impacts their geographic-environmental contexts. Next, I examine how climate change challenges contractarianism and undermine 'the legitimacy of political authority'[17]. I term these challenges *state failures*, since nation states have failed to address the tensions triggered by climate change and migration. Against this background, there are also arguments 'in favor of open futures. These nonfiction works are society's most important forces, which may provide the fuel for our collective future vitality and provide us the raw materials from which to construct our ecological systems of the future.

---

15 Ian Ward, *Law and Literature: Possibilities and Perspectives*. (Cambridge: Cambridge University Press, 1995).

16 For more on climate-change pop-science see Matteo Nicolini, "Law, the Humanities and Political Incertitude in a Time of Climate Change", *Legalities. The Australian and New Zealand Journal of Law and Society* 1.1 (2021), 91–115.

17 Ann Cudd and Seena Eftekhari, "Contractarianism", *The Stanford Encyclopedia of Philosophy* (Summer 2018 Edition), ed. Edward N. Zalta, 1. Available online at: https://plato.stanford.edu/archives/sum2018/entries/contractarianism/ (accessed 7 January 2023).

In opening up our future in a time of climate-induced changes, these forces stir the formalistic legal approach to climate change and make the law act as a bridge linking our troubled 'reality to an imagined alternative,' that is, an open and inclusive future.[18]

Global Warming and the Melting Bonds of Political Communities

All our non-fictional texts share a common feature: they address the insurgent ecological crisis in a distressing style, something after the fashion of an apocalyptic lexicon. *The Uninhabitable Earth* points to 'climate horrors,' 'sufferings,' and 'deprivation' (*UE*, 198, 200 and 192). *This is not a Drill*[19] is oversaturated with either the term 'extinction' or ones of its associated lexical items (*ND*, 7, 30, 78). *Inheritors of the Earth* warns us against a 'human-created mass extinction' (*IE*, 117; see also *UE*, 173). Finally, the books are percolated through by the ideas of 'climate and ecological emergency' and 'collapse' (*ND*, 9; *LE*, 5; *UE* 155; *PW*, 19).

Moreover, *The Uninhabitable Earth* points to the 'ecological and political devastation' prompted by climate change (*UE*, 172). Climate change, it maintains, 'promises to transform everything we thought we knew about nature, including the moral infrastructure of those tales' (*UE*, 150). In the midst of the environmental crisis, we perceive a profound sense of 'annihilation.' Sociologists designate this feeling 'Sociology of Loss.'[20] We are indeed experiencing the 'passing away' of our environmental and political tales 'characterized by open markets that ensure (sustainable) development and (environmentally-friendly) prosperity.'[21]

The idea of our political and ecological 'passing away' is linguistically and culturally embedded in the Western world. According to the *OED*, 'to pass away' means 'to perish or disappear; to be dissolved, to cease to exist.'[22] The breakdown of our physical heavens makes the 'relationship between human mobility and cli-

---

18 Alan Watson, *Failures of the Legal Imagination.* (Philadelphia: University of Pennsylvania Press, 1988), 36.

19 Extinction Rebellion, *This Is Not A Drill. An Extinction Rebellion Handbook.* (London: Penguin, 2019). Further references in the text, abbreviated as *ND*.

20 Rebecca Elliott, "The Sociology of Climate Change as a Sociology of Loss", *European Journal of Sociology* 59.3 (2018): 301–337.

21 Xifaras, "The *Global Turn*", 216.

22 s.v. "to pass away" Phrasal Verb 1, in *OED* https://www.oed.com/view/Entry/138429?redirectedFrom=pass+away#eid31648223 (accessed 13 January 2023).

mate change ... profoundly geographical.'[23] Firstly, the relation is geographical because it is earth dependent: evidently, 'There is No Planet B.'[24] The chapter headings of *The Uninhabitable Earth* are revealing: parts of our earthly environments will be rendered un-survivable by 'Heat Death,' 'Dying Oceans,' 'Freshwater Drain,' 'Unbreathable Air,' and 'Plagues of Warming.' Secondly, our 'legal structures and practices' are shaped by 'the material experience of concrete spaces and environments:'[25] politico-legal arrangements presuppose geographic scenarios. For example, several towns have been defined by 'waterways, and 'watery edges'.[26] Owing to rising sea levels, 'water has become a destructive as well as a functional and a scenic material,' whose 'malign threat' will affect them. 'Built at the edge of the seas,' coastal cities like New York, Mumbai, or Rio will 'risk episodic flooding' – if not be completely underwater (*BD*, 272). Likewise, small islands 'face the risk of becoming practically uninhabitable, and 'entirely underwater, with economic consequences that point to both their submersion and the 'loss of [their] territory.'[27]

Thirdly, we are experiencing the collapse of our political earth and heaven. This entails the demise of the legal arrangements whereby we tried to restrain carbon emissions. The 1992 UN Framework Convention on Climate Change and the 2015 Paris Agreement seem to encapsulate (or are underpinned by) a Judeo-Christian sentiment. Both were supposed to be our *katechon*, 'what withhold that' (2 *Th.* 2:6) global average temperature be increased, and what makes efforts to limit it 'well below 2 °C above pre-industrial levels' at least.[28] When it comes to curbing carbon emissions, though, 'the mystery of iniquity doth already work' (2 *Th.* 2:7). We have progressively removed restraints to emissions. 'Influenced by corporate lobbyists,' the signatories to both the 1997 Kyoto Protocol (and then the 2009 Copenhagen Accord) managed to establish an emissions trading system, thus commodifying carbon dioxide, i.e. the principal greenhouse gas. This legal arrangement favors the 'big emitters', that is, nations which are the largest CO2 con-

---

23 Baldwin, "Pluralising climate change", 516.
24 Mike Berners-Lee, *There Is No Planet B: A Handbook for the Make or Break Years* (Cambridge: Cambridge University Press, 2019).
25 Manderson and van Rijswijk, "Introduction", 168.
26 Richard Sennett, *Building and Dwelling. Ethics for the City* (London: Penguin, 2019), 271. Further references in the text, abbreviated as *BD*.
27 Rafael Leal-Arcas, *Climate Change and International Trade* (Cheltenham: Edward Elgar, 2013), 46 and 51.
28 Art. 2.1(a) of the Paris Agreement adopted under the UNFCCC in FCCC/CP/2015/10/Add.1, decision 1/CP.21. See Lavanya Rajamani, "The 2015 Paris Agreement: Interplay between Hard, Soft and Non-Obligations", *Journal of Environmental Law* 28.2 (2016): 337–358.

tributors.[29] But it also favors non-country actors, such as 'international aviation and shipping:' although they are big emitters, they are located outside the scope of the Paris Agreement.[30] Failures like these cause what *Climate Leviathan* denotes the 'principle failure':

> [The] Paris Agreement does not keep fossil fuels in the ground, but this does not mean it will not set the foundation for adaptation on a burning planet. On the contrary, the so-called '"failures' of Paris are enabling, and part of, a crucial adaptation, the adaptation of the political. (*CL*, 38)

This also explains why the 2021 draft of the Glasgow Climate Pact and 2022 the Sharm el-Sheikh Implementation Plan 'coal down, our hopes of limiting greenhouse emissions.[31] Instead of limiting emissions from private actors, our democratic societies have endorsed 'Climate Behemoth' (*CL*, 45) i.e. the 'mutual support for capitalism and for the nation-state' (*UE*, 192). In order to protect its own interests, capitalism 'overruns the world's borders to address the planetary crisis' (*UE*, 192), and seeks a connection with both the public sphere and transnational communities. Capitalism usually draws this connection by turning its interests into *the* organizing theme of political communities. Since these now count as the interests of the whole, our political bonds have been led astray by what transnational private actors have promised to us: 'Commerce would civilize manners; offer everyone the benefits of peace, human rights, representative democracy, and moderate government.'[32]

However, a new 'geopolitical struggle' has begun; in order 'to control the flow of resources,' mainly fossil fuel energy, 'from and through the north,' hence, 'capitalist states ... address the problems they have created by deepening' them (*CL*, 8). Furthermore, this new acquisitive organizing principle has prevented us from realizing, *inter alia*, 'that several small island nations ... face the risk of becoming practically uninhabitable,' 'economically nonviable,' and of being 'entirely underwater'

---

29 Art. 17 of the 1997 Kyoto Protocol. On such commodification see Michael Watts, 'Commodities,' in Paul Cloke et al. (eds.), *Introducing Human Geograhies* (3d edn, Adingdon: Routledge, 2014): 391–412, 406–407.
30 On big emitters, aviation and shipping see Alice Larkin et al., "What if negative emission technologies fail at scale? Implications of the Paris Agreement for big emitting nations", *Climate Policy* 18.6 (2018): 690–714, 692.
31 See Glasgow Climate Pact, Part IV (Mitigation) Nos 17 ('rapid, deep, and sustained reductions in global greenhouse emissions') and 20 ('accelerating efforts towards the phase-out of unabated coal power and inefficient fossil fuel subsidies'). See also Matteo Nicolini, Lega Geography. Comparative Law and the Production of Space (Cham: Springer, 2022), 252–255.
32 Xifaras, "The *Global Turn*", 216.

This has both economic- and human-related consequences. The economic consequences point to both their submersion and the 'loss of [their] territory.' Certainly, this will affect states' rights to a marine territory under Article 76 of the UN Convention on the Law of the Sea (UNCLOS)[33]. We do not still know, however, whether the UNCLOS considers the boundaries of EEZs as permanent. To put it another way, it is disputed whether the submersion of the islands – such as the Maldives (*ND*, 31–34) – would unleash their former seabed to the pillaging from economic transnational actors, which may try again to convince us of an unceasing economic growth.

The human-related consequence is even more distressing. The loss of territories will trigger impressive migrations. Rising sea levels would indeed 'made [them] to err from' their lands (*Is.* 63:17). These events will probably boost the 'movement of people across territorial borders, the mixing of bodies and places, and the reconfiguration of labor markets.'[34]

Climate Change and State Political Failures: The 'Constitutional Violation'

One of the winning arguments deployed by states is not to address climate change issues. They often exhibit *relaxed political tactics* towards it with the aim of preserving the capitalism-nation-state nexus and limiting "climate liability – for oil companies, for governments, for nations" (*UE*, 168).

In *The Code of Capital* (2019), Katharina Pistor explains why this occurs. Economic globalization holds a particular fascination *also* for states, which 'are not neutral when it comes to whose interests in an asset shall be given priority.' The prospect of benefitting from capital gains is 'more likely to find their blessings than claims that assert self-governance or seek to ensure environmental sustainability.'[35] To preserve the capitalism-nation-state nexus, we must carry on purchasing and consuming. Even on the brink of ecological collapse. Consequently, climate change and its disturbing arguments must be kept out of our sight. *Too much sen-*

---

33 Leal-Arcas, "Climate Migrants", 89.

34 Baldwin, "Pluralising climate change", 516. See Mathias Risse, "The Right to Relocation: Disappearing Island Nations and Common Ownership of the Earth", *Ethics & International Affairs* 23. 3 (2009): 281–300. Margaret Moore, *A Political Theory of Territory* (Oxford: Oxford University Press, 2015), 210.

35 Katharina Pistor, *The Code of Capital. How the Law Creates Wealth and Inequality.* (Princeton; Princeton University Press, 2019), 23.

sate, deep knowledge and conscious understanding of the catastrophe may cause a drop in commerce, threaten the collection of commodities, and reduce profits.

'Normal politic has failed us,' the authors of *This is Not a Drill* maintain, because granting 'access to government by big business ... has brought the whole planet to the brink of ecological disaster' (*ND*, 22–23). From a contractarian perspective, this challenges state authority because of its failure to tackle climate-change induced migration, as well as to manage the tensions that arise from environmental issues. Despite this, however, the 'Climate Behemoth' is still fascinating for those to whom globalization promises an 'impeding radiant future.'[36]

States have adopted the same organizing theme of global capitalism, which deliberately sacrifices the interests of political communities in favor of the gains of a limited elite. Not only does this raise concerns about the legitimacy of state policies, but it also questions the authority of the global actors in challenging political obligations. As noticed, this has triggered a change in how societies perceive their political bonds: interests are now arranged after a hierarchical scale, thus causing an imbalance between the interests of the communities and the interests of the economic actors. In case of conflict, it becomes clear, the interests of the former prevail.

Within contractarianism, this order of interests raises concerns as to the legitimacy of the current political obligation. In order to support the economic interests of its economic sectionalities, sovereign states have ignored the geographical and socio-cultural contexts within which the bonds of communities and the mobility-climate change nexus take place. By acting selfishly, then, we accepted the idea that 'economic growth [would] save us from anything and everything' (*UE*, 115) – global warming included.

When global actors control the levers of wealth distribution, the availability of resources is even more reduced, and 'our fragile web of life ... poisoned and broken' (*ND*, 6). This means that the political obligation becomes unsustainable. In times of political and environmental crisis, the conveyancing of the political obligation should seek an equitable balance of bargain powers and conflicting interests. *This is Not a Drill* suggests 'Every parliament, state legislature and local authority [need] to declare a climate and ecological emergency,' following the lead of more than 18 countries, the EU, and more than 2,000 local councils throughout the world (*ND*, 22). The UK House of Commons was the first House to declare:

> an environment and climate emergency following the finding of the Inter-governmental Panel
> on Climate Change that to avoid a more than 1.5 °C rise in global warming, global emissions

---

36 Xifaras, "The Global Turn", 216.

would need to fall by around 45 per cent from 2010 levels by 2030, reaching net zero by around 2050 ...[37]

Several other countries soon ensued.[38]

The capitalism-state nexus has marginalized climate-change issues in almost all political discussions. This hinders a further failure: 'that of state authorities to create the law that was wanted' by the political community, i. e. a law which should be firmly rooted in an environmentally, sustainable egalitarian commitment. The 'readiness on the part of the state authorities to allow' the private international investors to 'make a considerable part of the law' discloses state successful opportunism towards these contentious topics.[39]

This probably explains why Extinction Rebellion has recently declared the same 'bonds of the social contract ... to be null and void:'

> When Government and the law fail to provide any assurance of adequate protection, as well as security for its people's well-being and the nation's future, it becomes the right of its citizens to seek redress in order to restore dutiful democracy and to secure the solutions needed to avert catastrophe and protect the future (*ND*, 1–2).

This resonates with Johannes Althusius's idea of gross 'constitutional violation,' which entails a general callout to the public when there is any 'egregious', 'chronic,' 'persistent,' 'pervasive,' 'willful,' 'intentional,' or 'widespread' 'breach of a ruler's constitutional duties.'[40] There are hints of *abusus potestatis publicae* both when a ruler 'violates, changes, overthrows, or destroys' the fundamental law and people's rights; and when, as Althusius indicates, the abuse disrupts 'the natural laws and rights' (i. e., the environmental context) on which the constitutional framework is based.

Against this background, we do not need what, in contractarian terms, Theodore Beza termed as the 'theory of self-defence,' whereby 'sorting out what people

---

37 HC Hansard 1 May 2019, vol. 659, Columns 317–318.

38 A comprehensive list of climate emergency declarations is available at https://climatee mergencydeclaration.org/climate-emergency-declarations-cover-15-million-citizens/ (accessed 7 January 2023).

39 Watson, *Failures*, 87, 96.

40 Johannes Althusius, *Dicaeologicae libri tres, totum et universum jus, quo utimur methodice complectentes*. Heidelberg 1617, I.113.9–17. The quotations from Althusius are from John Witte, Jr., *The Reformation of Rights. Law, Religion, and Human Rights in Early Modern Calvinism*. (Cambridge: Cambridge University Press, 2007), 201.

could and should do when a political structure [goes] awry.'[41] The manifold lawsuits brought against oil companies and states on the grounds of climate liability count, for our purposes, as forms of self-defence. Certainly, climate-liability lawsuits are highly productive when it comes to challenging the Behemoth connection; and yet, they operate within traditional precincts and resort to the same formalistic legal arguments which saturate the capitalism-state nation nexus and its outcomes.[42]

Within their own precincts, states may exhibit *relaxed political tactics* and therefore play the game of 'strategic formalism.' I understand that this type of formalism may 'camouflage law' and deny 'the political moral, social choices which should be involved in any legal decision making even in hard cases,' such as climate change. This also accounts for another pattern of strategic formalism: the lack of pieces of legislation directly related to climate change amplify 'the disparate perspectives of scientist and lawyers' in this ambit and result in 'tensions and disagreements about how to use the law.'[43] In so doing, states hide their own agenda, i.e. the preservation of the capitalism-nation state nexus,[44] and 'conspire to limit climate liability – for oil companies, for governments, for nations' (*UE*, 168).

Migration and Climate Change: Colonial Hierarchies and their 'Unnatural' Arguments

Strategic formalism causes politico-ecological failure when the bonds 'of trust, loyalty, protection, and assistance between the citizen and the state' are severed – or, at least, harmfully shaken.[45] As it backs the capital-state connection, strategic for-

---

41 Theodore Beza, *Letter to Bullinger* (December 1574) as discussed in Witte, Jr., *The Reformation of Rights*, 124.
42 Novel "Climate Change"; Catriona McKinnon, "Climate crimes must be brought to justice", *The UNESCO Courier*, 3 (2019): 10–12.
43 Braverman, *Coral Whisperers*, 163.
44 Zdeněk Kühn et al., "EU law and Central European judges: Administrative judiciaries in the Czech Republic, Hungary and Poland Ten years after Accession". In Michal Bobek (ed.), *Central European Judges Under the European Influence: The Transformative Power of the EU Revisited.* (Oxford: Hart, 2015), 43–72, 45.
45 Andrew Shacknove, "Who Is a Refugee?" *Ethics* 95.2 (1985): 274–284, 279. See also Matthew Scott, *Climate Change, Disasters, and the Refugee Convention.* (Cambridge: Cambridge University Press, 2020, 4.

malism also is an enthralling organising theme for state communities. This is apparent when it comes to managing climate-induced migration.

This is apparent as regards migration. The global supply chain needs labour to work. Capital uses several arguments to convince prospective workers to migrate to the places in which it produces. These arguments are convincing since they give migrants the illusion of taking part in the global distribution of workers, who are attracted by the pursuit of happiness and the perspective of improving their (and their family's) standard of life. The global supply chain mobilizes mass migration from the Global South towards the Global North with a 'promise of growth [that] has been the justification for inequality, injustice, and exploitation,' with more wounds to heal 'in the near climate future'(*UE*, 166).

Under the aegis of the UN, two Global Compacts were signed in 2018, namely the *Global Compact for Safe, Orderly and Regular Migration* (GCM), and the *Global Compact on Refugees* (GCR)[46]. The GMC acknowledges that migration has been 'part of the human experience throughout history;' it also refers to migration as 'a source of prosperity whose 'positive impacts can be optimized by improving migration governance.' Furthermore, the signatories to the compacts acknowledge that both instruments are 'complementary international cooperation frameworks,' because 'migrants and refugees may face many common challenges and similar vulnerabilities.'[47]

There is something treacherous in such sanctification of global mobility. Climate-induced refugeeism, indeed, hardly squares with existing legal instruments, such as the 1951 Geneva Refugee Convention and the 1969 OAU Convention Governing the Specific Aspects of Refugee Problems in Africa.[48] According to the High Court of Australia, for example, 'a person fleeing from' natural disasters does not qualify as refugee under the 1951 Convention.[49] The UN High Commissioner for Refugees (UNHCR) does the same, because the 1951 Convention 'rules out such persons as victims of famine or natural disaster, unless they also have

46 Resolution adopted by the General Assembly on 19 December 2018, UN Doc A/RES/73/195 (11 January 2019) Annex: Global Compact for Safe, Orderly and Regular Migration. Report of the United Nations High Commissioner for Refugees (13 September 2018) UN Doc A/RES/73/12 (Part II) Global Compact on Refugees.
47 Paras 3 and 4 of GCM. See Elizabeth G. Ferris, Katharine M. Donato, *Refugees, Migration and Global Governance: Negotiating the Global Compacts.* (Abingdon: Routledge, 2020), ch 5.
48 Christel Cournil, "The inadequacy of international refugee law in response to environmental migration". In Benoît Mayer and François Crépeau (eds.), *Research Handbook on Climate Change, Migration and the Law.* (Edward Elgar: Cheltenham and Northampton, MC, 2017), 85–107. On the OAU Convention see Rafael Leal-Arcas, "Climate Migrants: Legal Options". *Procedia – Social and Behavioral Sciences* 37 (2012): 86–96, 94.
49 *Applicant A v Minister for Immigration and Ethnic Affairs* [1997] HCA 4, [1997] 190 CLR 225.

well-founded fear of persecution.'[50] Nor is the 1967 Protocol applicable, since it does not cover 'sociopolitical factors,' such as 'fleeing climate breakdown' or 'economic collapse' (*HE*, 31).[51] A citizen from Kiribati applied to New Zealand in order to qualify as the 'first' climate change refugee in 2015. There was 'no evidence,' the Supreme Court argued, 'that the Government of Kiribati [was] failing to take steps to protect its citizens from the effects of environmental degradation to the extent that it can.'[52] That is why, in 2015, the International Organization for Migration (IOM) addressed the topic establishing a dedicated 'Migration, Environment and Climate Change' (MECC) Division.

Legal scholars defend such disqualifications using strategic formalism. Formalistic arguments are indeed rooted in the principle of non-discrimination as *the* central tenet of refugee law. Climate-change related 'disasters are not considered "political" events.' Certainly, they are 'sources of vulnerability;' their being 'beyond social control,' however, imposes 'no obligation on a government to secure a remedy.'[53] In addition, persecution or discrimination are 'integral [features] of the refugee definition' under international law.[54] Quite ironically, climate change has a democratic allure, because it hammers without discriminating among individuals.

Furthermore, as its para 7 affirms, the GMC has 'a non-legally-binding' character and is not enforceable. This minimizes the extent to which 'Natural disasters, the adverse effects of climate change, and environmental degradation' have to be tackled as regards migration. In addition, the narrative of prosperity and progress hides a *narrative of superiority*, where the 'radiant future' covers an imbalanced correlation between the actors of globalization. Egalitarian in its commitment, this narrative facilitates the harmonization and the convergence of laws in order to stimulate business and economic development. It also proposes a transnational, borderless legal framework into which heterogeneous legal systems coalesce. The pursuit of happiness and the illusion of taking part in the global distribution of wealth mobilizes mass migration. As *The Uninhabitable Earth* upholds,

---

50 UNHCR, *Handbook on Procedures and Criteria for Determining Refugee Status and Guidelines on International Protection Under the 1951 Convention and the 1967 Protocol Relating to the Status of Refugees*, HCR/1P/4/Eng/Rev. 4 (2019), 39.
51 See *Matter of Acosta*, A-24159781, United States Board of Immigration Appeals, 1 March 1985. Available online at: https://www.refworld.org/cases,USA_BIA,3ae6b6b910.html (accessed 7 January 2023).
52 *Teitiota v Ministry of Business Innovation and Employment* [2015] NZSC 107 (20 July 2015), at 13.
53 Shacknove, "Who Is a Refugee?", 275.
54 Scott, *Climate Change*, 4.

'the promise of growth has been the justification for inequality, justice, and exploitation,' with more wounds to heal 'in the near climate future' (*UE*, 166).

When economic migrants legally (and sometimes illegally) cross the fences of our *hortus conclusi*, they enter the walled garden, into which (as well as into its supply chain) they become integrated. Our notion of integration, though, is economical, not ethical. It is functional only to production and consumption of commodities.

What do these migrants expect from us? Integrated in our safe spaces economically, they must comply with the following rules: they must take part in the supply chain; like us, they must spend their stipends (their share in the global wealth) on the commodities they produce to sustain the economic cycle; finally, after having worked in our safer places, they must retreat from them. The last requirement is also known as the NIMBY (Not In My Back Yard) factor, which reflects what has been termed as the 'Global Apartheid.'[55] As consumers, shareholders, and entrepreneurs, we turn workers into *something servient* to the comfort zones of capitalist economies. Again, our motto revolves around the maximization of financial returns and the idea of the safe space: the quieter the space, the safer it is, and the higher the returns are.

In his *Reimagining Britain*, the Archbishop of Canterbury Justin Welby highlights how economic integration makes communities (even parish communities) 'resistant to outsiders and reluctant to sponsor diversity if there is any possibility that it will diminish housing values or be perceived as a risk to the aesthetics of that community.' There is no 'diversity, or mixing, no valuing of difference, no sight of the poor'; this sharply contrasts with the biblical (and Christian) mandate.[56]

Ultimately, gentrification recreates a safe space whose community is homogenous, 'expensive,' and 'exclusive.' In Welby's words, 'the balance of property values and influence of location' are extremely 'materialistic' and point to the 'economic maximization' of investments in housing.[57] The presence of outsiders makes the space diverse and no longer exclusive, that is, less expensive if housing prices drop because of the *mixité*. That is why Western states have reproduced the 'racially marginalized hierarchies' between the metropolis and its colonies within their urban areas. As a result, migrants are located into cordoned-off 'periphractic

---

55 Yajaira Ceciliano-Navarro, Tanya Gloash-Boza and Luis Rubén González Márquez, "Reflections on Anti-Immigration Narratives and the Establishment of Global Apartheid". In Molly Katrina Land et al. (eds.), *Beyond Borders. The Human Rights of Non-Citizens at Home and Abroad.* (Cambridge: Cambridge University Press, 2021), 94–110.
56 Justin Welby, *Reimagining Britain. Foundations for Hope.* (London: Bloomsbury, 2018), 145.
57 Welby, *Reimagining Britain*, 146.

spaces,' which constrain immigrants 'in terms of location and their limitation in terms of access – to power, to (the realization of) rights, and to goods and serv-ices.'[58]

Displacement and spatial marginalization trigger a 'really hostile environment for [allegedly] illegal immigrants' (*HE*, 2),[59] i.e. a by-product of 'heightened border measures that restrict migration, measures that play well with anxious electorates and hawkish politicians.' It might be argued that, in so doing, states disavow 'the West's complicity in the wider social, political and economic conditions that con-tribute to the migrations the West seeks to secure.'[60] But this also accounts for an-other application of strategic formalism: the securitisation of climate-induced mi-gration. Whilst praising the economic benefits of global mobility and the protection of migrants' rights, 'politicians work together across borders to make it more dif-ficult for people to move, while capital is allowed to flow freely' (*HE*, 35).

Take the EU policies related to migration and refugeeism. On the one hand, they are strongly inclusive but, on the other hand, EU treaties and legislation are overpopulated with references to external borders. Schemes such as Frontex are 'the culmination of the process of securitization of the EU external borders,' whereas the European Border Management (EBM) has 'multiple dimensions': it en-compasses several methods of 'border checks and surveillance,' which make it hard for migrants to access the EU.[61] Indeed, EU citizenship complements national citizenship, which means that third-country nationals residing within its borders are, in principle, excluded from it. To make the EU a safer space, it has been 'nec-essary to draw a line between insiders and outsiders, between "us" and "them".'[62] What is walled is not the border, but the space where the integration process takes place; the EU thus establishes 'a unitary basis for exclusion rather than a coherent set of criteria for inclusion.'[63] Evidently, the 'principle of solidarity and fair sharing of responsibility,' which underpins the Area of Freedom, Security, and Justice laid down by the Treaty on the Functioning of the EU, is a privilege reserved for EU citi-

---

58 David Theo Goldberg "'Polluting the Body Politic': Race and Urban Location". In Nicholas Blom-ley et al. (eds.), *The Legal Geographies Reader: Law, Powers, and Space.* (Oxford: Blackwell, 2001), 69–76, 71, 72.

59 The so-called 'hostile environment immigration' policy was coined by the then UK Home Sec-retary Theresa May. See James Kirkup and Robert Winnett, "Theresa May interview: "We're going to give illegal migrants a really hostile reception"", (London) *The Telegraph*, 25 May 2012.

60 Baldwin, "Pluralising", 521..

61 On the so-called "Fortress Europe" see David Delaney, *The Spatial, the Legal and the Pragmatics of World-Making.* (Oxford: Blackwell, 2010), 153.

62 Serhii Lashyn, "The Aporia of EU Citizenship". *Liverpool Law Review* 42 (2021): 361–377, 364.

63 Jaqueline Bhabha, "'Get Back to Where You Once Belonged': Identity, Citizenship, and Exclusion in Europe". *Human Rights Quarterly* 20 (1998): 592–627, 604.

zens, whereas third-country nationals must merely be treated fairly when they try to access the EU.[64]

The same holds true as far as the litigation for the erection of the U.S.-Mexico border wall is concerned. When granting President Trump the petition for a writ of certiorari for the erection of the wall, the U.S. Supreme Court explicitly ranked public-finance concerns as superior as regards ecologic and solidarity issues. In the Court's reasoning the 'construction of a border barrier ... would cause irreparable harm to the environment and to [individuals;' however, if the federal government were 'unable to finalize the contracts [for the wall] then the funds at issue will be returned to the Treasury' – and cost borne by the U.S. taxpayers. Such impressive use of strategic formalism 'is a straightforward way to avoid harm to both the Government and respondents while allowing the litigation to proceed.'[65].

Included but marginalized, migrants enter irredeemable conveyances with the capitalism-nation state nexus and bear the costs of politico-legal marginalization. This creates a flaw within the contractarian speculation. Although 'Nature herself seems to proclaim this with a loud voice,' i.e. that 'rulers receive their authority ... by the free and lawful consent of the people,'[66] it is undeniable that the capitalism-state nexus arranges the political bonds of the community after 'structures of [economic] authority and [political] obedience.'[67] Althusius termed these bonds 'unnatural.' Migrants – even the climate-induced ones – 'subject themselves' to this organizing principle and accept its 'authority by their own consent and voluntary act.' Within our wealthy communities, they 'subject themselves to these "unnatural" structures and strictures of authority, for they realize that without them ... even their most basic rights will mean little.'[68]

---

64 Article 80 TFEU. See Violeta Moreno-Lax, *Accessing Asylum in Europe: Extraterritorial Border Controls and Refugee Rights under EU Law.* (Oxford: Oxford UP, 2017), 31, 34. See also Andrew Burridge et al., "Polymorphic borders", *Territory, Politics, Governance,* 5 (2017): 239–251, 242.

65 U.S. Supreme Court, *Donald J. Trump, President of the United States, et al. v Sierra Club, et al.* on application for stay 588 U.S. (2019) 1, 26 July 2019.

66 François Hotman, "Francogallia". In Julian Franklin (ed.), *Constitutionalism and Resistance in the Sixteenth Century: Three Treatises by Hotman, Beza, and Mornay.* (New York: Pegasus, 1969), 55–70.

67 Witte, Jr., *The Reformation,* 183.

68 Witte, Jr., *The Reformation,* 184.

The Force of Arguments and Our Egalitarian Misery

There is a further argument that the capital-state nexus uses to convince us that we still live in safe places. The force of this argument lies in its egalitarianism and universalism: every community must confront climate change, which we can tackle by adopting a global-response strategy. This response is supplied by global capital, which operates on a transnational scale. To favour the smooth flow of goods and services across frontiers and national legal systems, global markets approach territories and borders through the application of generalizations and "global-business-as-usual" strategies.

Yet, these strategies are unlikely to 'produce predictable outcomes as laboratory experiments might.'[69] Like cross-border economic interests, ecological concerns are also shared by transnational communities, but, unlike economic interests, our concerns are extremely concrete. Beyond global law, there are communities. Climate change is certainly an international, cross-border, and domestic phenomenon. As a transnational phenomenon, it transcends borders, but it is also a concrete concern, striking indiscriminately on a territorial basis.

Global-business-as-usual strategies lead us to fall prey to Hayek's 'synoptic delusion.' It certainly gives us the illusion that, from a central standpoint, we can access all the information we need to respond effectively to our ecological problems. In Hayek's opinion, though, this is the 'characteristic error' of those who underestimate 'concrete facts,' such as the complexities of our environmental crisis. Besides this, the 'synoptic delusion' replaces the social reality we seek to regulate with 'a surveyable whole of all data' and arguments complemented by the capital-state nexus. This prevents us from tackling climate change with a more nuanced and contextual approach, leaving us intoxicated by 'the sense of unlimited powe'" that arises from the global strategy of ecological adaptation.[70]

Such a synoptic delusion, I assume, reflects a kind of egalitarian misery which is the paradigm of adaptation.[71] We have deceived ourselves. Bracketed, as we are,

69 Cinnamon Carlarne and Daniel Farber, "Law beyond Borders: Transnational Responses to Global Environmental Issues". *Transnational Environmental Law* 1.1. (2012): 13–21, 19.

70 See Friedrich A. Hayek, *Law, Legislation and Liberty*, Vol I, *Rules and Order* (London: Routledge & Kegan Paul, 1973), respectively 14, 8, and 14–155. I owe this set of reflections on Hayek's predicaments to Richard Mullender. See Richard Mullender, "Negligence, Public Bodies, and Ruthlessness", *Modern Law Review* 72.6 (2009): 961–983, 973.

71 On the concept of "egalitarian misery" see Richard Mullender, "On the French Revolution and the Programmatic Imagination: Hilary Mantel on Law, Politics, and Misery". In Richard Mullender,

within the winning arguments of the nation-state, we accept that adaptation is both the winning argument of capitalism and the 'progress' of our time. We know that the world is on fire. To be honest, we really do not care a straw about it. The decline is a misery; it sounds so egalitarian because it consumes all our lives in our safe spaces. Yet, the wealthy will decline in better spirits than the lowly and marginalized.

Being the 'Authors of What Comes Next:' A New 'Sustainable,' Organising Political Principle in a Time of Climate Change

Several conclusions may be drawn on how strategic formalism is applied to the current ecological crisis. Firstly, its application encourages the recourse of a formalistic approach when it comes to managing climate change. The law has traditionally addressed climate change and migration through an approach which advocates the application of traditional legal devices that international, supranational, and domestic orders make available to us. However, this approach scarcely squares with the current state of affairs. Under no circumstances do climate migration and refugeeism qualify under the international law of refugee protection, but strategic formalism also does not formulate innovative responses to the concerns raised by climate change. Take, for example, climate litigation, especially the lawsuits filed against carbon-emitters.[72] Strategic formalism tends to adapt them to traditional taxonomies, such as climate liability, thus normalising the same idea of ecological and climate loss. This also holds true for the Warsaw International Mechanism for Loss and Damage associated with Climate Change Impacts (WIM). While implementing Article 8 of the Paris Agreement, its role is limited to 'recommendation and assistance' for climate-related losses.[73]

Matteo Nicolini, Thomas DC Bennett and Emilia Mickiewicz (eds.), *Law and Imagination in Troubled Times: A Legal and Literary Discourse.* (Abingdon: Routledge, 2020): 133–156, 153.

72 See e.g. Francesco Sindico et al. (eds.), *Comparative Climate Change Litigation: Beyond the Usual Suspects.* (Cham, Springer, 2021).

73 Decision 2/CP.19 "Warsaw international mechanism for loss and damage associated with climate change impacts", Report of the Conference of the Parties on its nineteenth session, held in Warsaw from 11 to 23 November 2013, FCCC/CP/2013/10/Add.1. See Frank Bierman and Ingrid Boas, "Towards a global governance system to protect climate migrants: taking stock". In Benoît Mayer and François Crépeau (eds.), *Research Handbook on Climate Change, Migration and the Law.* (Edward Elgar: Cheltenham and Northampton, MC, 2017), 405–419, 410.

The second conclusion regards how scholars should tackle migration and environmental issues. As a comparative legal scholar, I understand that we have to go above and beyond the boundaries marked by strategic formalism. This, it becomes clear, entails reframing the political bonds of our communities and considering our 'future as a site of infinite potential rather than foreclosure.' Bypassing strategic formalism also entails expanding our political-legal imagination so that 'we can invent the future we want, rather than merely prepare for the future' that the capitalism-state nexus and its 'experts tell us we should expect.'[74]

That is why I have constantly engaged in a conversation with non-fictional texts related to climate-change issues. If they represent the output of the most active forces within our society, we must point to them when stretching our thinking into the future. Letters to the Earth invites us to 'be the authors of what comes next': we should mobilize our societies and their forces because we need 'the largest creative response to these times of crisis the world has yet seen' *(LE, 6)*. This is Not a Drill advocates for the 'liberation of our minds from colonizing categories' *(ND, 7)*, which may also entail 'undertaking mass civil disobedience to create a new political reality the whole world over' (ND, 22). The Uninhabitable Earth points to what Mann and Wainwright indicate as the 'Climate X', that is, 'a global alliance operating in the name of a common humanity' *(CL, 173)*, rather than in the interest of its sectionalities, such as "the interests of capital or nations" *(UE, 192)*.

I am not advocating the 'creation of spaces of collective rebellion' but, as a legal scholar, I encourage 'every principled and peaceful citizen' to challenge the application of strategic formalistic approaches, to liberate ourselves from the economic categories we have been imposed by the 'Climate Behemoth' connection, and therefore to arrange political arrangements around new organising principles.

The third conclusion relates to the methodological approach we have applied when tackling migration in a time of climate change. The cross-disciplinary method that law and the humanities provide has proved to be a highly imaginative activity in itself, since it has 'the potential to broaden and deepen the individual's [as well as the collective] understanding of ethics, politics, and human relations.'[75]

It is not just a matter of upholding climate justice or addressing 'the distributive effects of climate change.'[76] When we open it up to contexts of mutual interference and interdisciplinarity, the law ceases to be a mere dry legal text and allows us to read it 'in a wider variety of ways' thus filling the gap 'between

---

74 Baldwin, "Pluralising climate change", 526.

75 James Seaton, "Law and Literature: Works, Criticism, and Theory". *Yale Journal of Law & the Humanities* 11.2 (1999): 479–507, 479, 580.

76 Joseph Wenta et al., "Enhancing Resilience and Justice in Climate Adaptation Laws". *Transnational Environmental Laws* 8.1 (2019): 89–118, 110.

general law and more perfect justice'. This also entails continually '[stirring] up what the law,' as well as its formalistic approach, sets down.[77] This is the kind of legal pragmatism which adequately reflects the transnational concerns underpinning the law and the new global insecurities triggered by climate change. That is the reason why, in order to stir the rules of global law, it is necessary to mobilize the most active forces within society.

When stirring the law, these groups give raise to a new organizing principle, which reflects 'the view that, as conceptions of justice change over time, so too should' our legal rules in a time of climate change.[78] Whilst bridging legal reality to its imagined alternative, I understand that the legal mobilization of the most active forces within societies perfectly matches 'the deep-rooted Anglo-American practice of the common law, according to which the law involves an unceasingly exercise of imagination.' As it has been vividly noticed, 'This can be partly due to the inherently open-textured nature of the common law. Those entrusted with the task of interpreting and reshaping the common law must remain imaginatively vigilant at all times.'[79] It is not an accident that 'stirring the law' is a typical Anglo-British legal device, a highly productive metaphor, which describes the relationship between equity and its formalistic counterpart, i.e., the more 'classical' common law as it had been molded in Westminster Hall.

The fourth conclusion considers the most suitable methodological approach in confronting the mobility-climatic nexus. To this extent, equity offers us a productive paradigm applicable to cross-disciplinary research in this ambit. How it stirs the law reminds us of its being 'a door having one side within the law and one side without,' whilst strategic formalism prefers' to keep the door closed and to see only their side of it.'[80] How equity assists us in redeeming the bonds, interests, and organizing themes within our communities is the same attitude which is required of 'every principled and peaceful citizen' to challenge strategic formalism incrementally.

---

77 Gary Watt, *Equity Stirring. The Story of Justice Beyond Law.* (Oxford: Hart, 2009), 1.
78 Richard Mullender, "Context, Contingency and the Law of Negligence (or from Islands to Islands of Time)". *Bracton Law Journal* 29 (1997): 23–33, 27.
79 Richard Mullender, Matteo Nicolini, Thomas DC Bennett and Emilia Mickiewicz, "Legal Imagination in Troubled Times. An Introduction". In Richard Mullender et al. (eds.), *Law and Imagination in Troubled Times: A Legal and Literary Discourse.* Abingdon: Routledge, 2022.
80 Watt, *Equity Stirring,* 1.

Subversive Legal Methodologies and Challenging the Narrative of Legal Decline

Comparative law has a potential as a discipline to engage in a methodological conversation with the cross-disciplinary ambit of research which is climate change. Because of its subversive and critical approach to the law, it runs somewhat counter the 'paradigm of biological decline' which is predominant 'among ecologists, environmentalists and conservationists' (*IE*, 30). This is not to deny the ecological threat: 'the extinction crisis is real,' *This is Not a Drill* warns us (*ND*, 30); as 'The world's climate is getting hotter, and they have nowhere to go', *The Inheritors of the Earth* concludes (*IE*, 76). But, as we have already seen, there are also hints of hopes, as well as arguments in favour of open futures. Notwithstanding the changes caused by anthropogenic drivers, biologists have demonstrated that, when copying with global warming, ecological biodiversity 'responds by evolving' (*IE*, 117), i.e. proposes a new biological organising theme whereby 'the biological diversity of the Earth' may be increased (*IE*, 117 and 9).

The same holds true in comparative legal studies. We should shackle off the 'pessimism-laden, loss-only view' of our future (*IE*, 117 and 9), which is reflected in how strategic formalism tackles with what we may term the legal biodiversity of the world. The 'Climate Behemoth' does, in fact, disregard legal diversity. Consequently, its transnational concern – the governance of both mobility and climate change along economic lines – should be confronted through homogeneous political-legal features throughout the world. Indeed, such features reflect the new economic organizing theme, which stimulates business, mobilizes mass migration and, at the same time, grants the illusion of taking part in the global distribution of wealth. In a time of climate change, however, it mainly manages (i.e., simplifies) legal complexity.

In a globalized economic world, strategic formalism assumes that the legal biological decline is consistent with the radiant future promised to humanity. This is an assumption that totally departs from what comparative legal studies presuppose, i.e., the sustainability of legal variety and legal diversity. Again, like equity, the mobilization of the most active forces does not mean breaking the rules, but merely bending them, by progressively introducing a set of 'comparatively novel legal principles' and therefore 'claiming to override the older jurisprudence,' which is strategic formalism and the organizing theme 'of the country on the strength of [its] intrinsic ethical superiority.'[81]

Arranging societal interests around novel organizing themes, such as climate justice and migration, may therefore increase the responsiveness of substantive laws to change in socio-ecological systems. Climate change and its induced migration should be confronted with a variety of legal responses. A Planet to Win, On Fire, and The Case for a Green New Deal advocate for a deal as the new foundation of our political bonds. On August 1, 2010, this entry was published.[82] Assisted, as we are, by unceasing exercises of legal imagination, we should indeed stir the law and explore it so that it might be interpreted and applied in a manner that responds to changing conditions.'[83] This form of pluralizing the legal debate may be extremely useful when it comes to addressing the distributive effects of climate change and contrasting the environmental racial segregation Western communities practice as regards migrants – even if they are climate-induced migrants. This might be fostered by 'new regulatory practices' and new 'forms of governance,' such as public participation or legislative frameworks whereby decision makers stir the law simply by adjusting 'existing legal processes to accommodate changing environmental or social conditions.'[84]

I would like to highlight a new organizing principle that, like equity, should be able to shake up the law and open the 'back door' to the diverse forces and interests of our communities. We do not require a law modified for climate change in a world altered by humans. We must investigate, through creative thinking and initiative, how to transform our transnational concerns about various global insecurities into new legal patterns, so that hostile ecological and political environments can eventually disappear.

---

81 Henry Sumner Maine, *Ancient Law. Its Connection with the Early History of Society, and its Relation to Modern Ideas.* (Cambridge: Cambridge University Press, 2012 [first published 1861]), 45.
82 Anne Pettifor, *A Planet to win. Why We Need a New Deal.* (London: Verso, 2019). Naomi Klein, *On Fire. The Burning Case for a Green New Deal.* (London: Penguin, 2019).
83 Wenta et al., "Enhancing Resilience", 108.
84 Wenta et al., "Enhancing Resilience", 113 ad 108, who quote the *Biodiversity Conservation Act 2016* (NSW), s. 1.3(d). See also Philippa C. McCormack, "The Legislative Challenge of Facilitating Climate Change Adaptation for Biodiversity". *Australian Law Journal* 92.7 (2018): 546–562.

Index

https://doi.org/10.1515/9783111081687-010

www.ingramcontent.com/pod-product-compliance
Lightning Source LLC
Chambersburg PA
CBHW062058270326
41931CB00013B/3134